THE
STRUCTURE OF STEEL
SIMPLY EXPLAINED

BY

ERIC N. SIMONS

AND

EDWIN GREGORY, Ph.D., M.Sc., F.I.C.

Author of "Metallurgy"
Chief Metallurgist, Park Gate Iron & Steel Co., Ltd.

With an Introduction by

F. C. LEA, O.B.E.

D.Sc., M.I.Mech.E., M.Inst.C.E.

WILDSIDE PRESS

First published February 1938
Reprinted April 1938

INTRODUCTION

By F. C. LEA, O.B.E.
D.Sc., M.I.Mech.E., M.Inst.C.E.

The process of steel making from wrought iron has been known for many centuries; but during the last eighty years, since Bessemer first made mild steel in the converter, there have been very remarkable developments, which have led to revolutions in engineering practice, and have made possible the construction of engines and machines, many types of electrical plant, and domestic and chemical utensils that would otherwise have been impossible.

Michael Faraday more than one hundred years ago made experiments on the alloying of certain other metals with iron. All the older steels depended for their properties upon the carbon content of the steel, and they continued to be made until Huntsman developed the method of making crucible steel, by melting bars at high temperatures, which had been carbonized in a manner somewhat similar to that in which case hardening is still practised.

Shortly after Bessemer produced steel in the converter, it was found that a certain percentage of manganese was necessary to remove oxygen from the molten steel and to overcome some of the difficulties due to the sulphur and phosphorus that were present. By increasing the manganese content and adding a considerable percentage of tungsten, Mushet produced the well-known "self hardening" steel, which can be said to be the forerunner of the modern high-speed steels.

In metallurgical works and laboratories all over the world very many experiments have been carried out during the last fifty years to determine the effect of various alloying elements, and to-day a very large number of alloy steels are being produced, which are being used in many industries, and have led to revolutions in industrial practice probably as great as any that have been brought about in any corresponding period in the world's history.

The discovery of the manganese steels by Sir Robert Hadfield, and of the stainless steels by Mr. Harry Brearley, the development of silicon steels and cobalt steels for the construction of magnets, the heat-resisting steels, the tungsten and tungsten-cobalt steels for tools and other purposes, have all had a remarkable influence upon modern engineering development.

In the chemical and dyeing industries the stainless and heat-resisting steels have made it possible to carry out industrial processes in vessels made of these steels, which before could only be carried out in other specially prepared vessels of less substantial and much more expensive materials. The motor-car and the aeroplane owe much to the development of suitable steels, capable of resisting the extreme conditions of service required in these remarkable modern aids to transport. In the workshop, lathes and other types of machines for removing metal have had, during the last thirty years, to be redesigned in order that they might use efficiently and effectively the remarkable tool steels that have been developed.

It is, however, of great significance and importance that nearly all the steels to which reference has been made, including the carbon steels, depend for their properties not only upon their composition, but also upon the particular treatments to which they are subjected during the course of manufacture, and their preparation for the purpose for which they are to be used.

It becomes increasingly desirable, therefore, that all those interested in steel and its use for any particular purpose should be aware of the very great importance of the particular " structure " that steel should have in order to meet particular conditions.

The book that Mr. Simons and Dr. Gregory have produced endeavours to supply in a simple way the information that should be available to anyone who is interested in any industrial problem in connexion with which some type of steel may be required to meet specific requirements.

The subject is important; but there are clearly many scientific and technical problems related to it which can only be adequately dealt with in large and specialized publications. This book only attempts to give what some may perhaps call a very superficial view of the subject of the structure of steel; but from the experience which the writer of this introduction has had amongst engineers and those engaged in other industries necessitating the use of steel, there seems no doubt that the matter contained in this book will be of value and interest to those who read it. One impression will no doubt be left in the mind of the reader— the desirability, yea, the necessity, of consulting the manufacturers as to the suitability of any one of the large number of steels available and the *particular heat treatment* that it should receive, in order that it may have the properties best suited to the particular purpose for which it is to be used.

.

PREFACE

The genesis of this book, viewed in far perspective, was the daily work of editing a technical journal on steel and engineering, the *Edgar Allen News*. Engaged in this from 1919 onwards, I had ample occasion to observe two significant facts. First, steel experts and metallurgists generally are convinced that what is common knowledge to them is common knowledge in the abstract, an assumption clearly untrue. Secondly, the people who buy, sell, study, manipulate, and in short have to do with, steel (with the exception of the experts aforementioned), thirst for an insight into steel and its structure, because they need this insight in their jobs and do not find it easy to come by.

In view of the spate of technical literature that issues from the printing presses of the world, this last statement may seem absurd. Nevertheless, examination shows that the literature on steel is either written by experts for experts—in which case the layman cannot make head or tail of it—or written by professors for students. But the academic textbooks (and I do not exclude from this even my collaborator's amazingly good book on *Metallurgy*) tend to assume in their readers a basis of chemical, practical, or scientific knowledge which not everyone possesses. Also they employ the simpler technical terms with a freedom that the ordinary person cannot emulate.

The position is further complicated by one or two queer kinks in human nature. On the one hand, certain experts regard the layman as an ignorant ass, and will not, therefore, go out of their way to make things clearer for him. Their attitude is the Yorkshire one of : " Tha knoas nowt abaht it, and tha'd best not meddle wi' what tha doesn't understand." On the other hand, the expert seldom writes well. Writing is a skilled job, and it is not easy for the unskilled labourer in this field to dig up and present his knowledge in digestible form, even when he is willing.

More and more, then, as the years went by, I became convinced that not only were there—as the correspondence files of the *Edgar Allen News* showed—hundreds of people all over the world craving for some simple and readily understandable treatise on steel and its structure, but also I myself, a non-technical man, was in exactly the same plight. I had been dealing with steel all my life, and it was still, for me, something " wropt in mystery ".

I toddled about among my colleagues for many moons, assiduously dropping seeds that I hoped would germinate in one or other of their minds and one day produce such a book as the present. But in the end I realized, as many a journalist has realized before me, that if one wants a book in everyday language on a particular subject, one has usually to write it oneself. So I set to work.

It was an audacious and rash thing for a non-technical man to do, but ignorance sometimes points the way for others to follow. I got hold of every textbook on steel I could find, and with a wet towel metaphorically round my head, and incredible pertinacity in asking tomfool questions of my busy but invariably genial and patient colleagues, the experts of Edgar Allen & Co., Ltd., put together a series of elementary articles on steel and its structure. I tried to understand the subject myself, and then, when I thought I had understood it, to express what I had gathered in a simple and perspicuous way. I made a free and often ill-advised use of analogy. I interrupted my narrative by interpretation and explanation wherever I thought the circumstances warranted it, and on the whole I enjoyed myself. My articles were published month by month in the *Edgar Allen News*, after being " vetted " to some extent by amused and kindly colleagues.

The success of the series was surprising, even to me. From all quarters poured in, not in ones and twos but in dozens, expressions of interest and appreciation, and requests for the entire series in book form.

Alas! there are pitfalls for the unwary in this simplification business, and as Dr. C. H. Desch (who took a lively interest in the articles) and other experts pointed out, I had fallen into some of them. It was obvious that here and there, in my laudable efforts to get down to rock bottom, I had descended too abruptly and landed with a disconcerting and painful bump. I had over-simplified in some places, exaggerated in others, and in one or two instances even misstated and misled. I do not apologize. Such errors were inevitable from the start.

It was soon obvious that if we were to republish the articles in book form, they would have to be worked over very thoroughly by someone with profound knowledge combined with skill in verbal presentation, someone who would retain as much of my work as was good and sound and eliminate all that was not.

Such a collaborator I found, to my delight, in Dr. E. Gregory. Dr. Gregory has restated and amplified, corrected and amended, my tentative articles (while retaining all their simplicity), so conscientiously and carefully that in many ways this is more his book than mine. He has turned a piece of popular journalism into a scholarly but still popular work. I confess to some sorrow at the ruthless way in which he has excised my more picturesque analogies. (On some of them I must secretly have rather prided myself.) But I believe that not only has he not weakened the book thereby, he has actually strengthened it.

To sum up, the scheme of this book and a proportion of the incidental writing are mine. The rest is all Dr. Gregory's, the last chapter in particular being wholly his. Our collaboration has been a courteous and happy one, and we are less interested in any credit that may attach to our book than in the fact that 'we have filled a serious gap in technical literature, and done, we believe, a good job of work.

For the first time, perhaps, the ordinary man, be he steel user, student, layman or worker, has here, in book form, the whole story of steel and its structure, told in words he can understand, told simply and straightforwardly, yet without inaccuracy or misstatement. In short, here is the book for Mr. Everyman who has to do with steel. We hope he will find it all he requires.

It would be ungrateful if I did not take this opportunity of thanking Messrs. L. K. Everitt, B.Met., and R. G. Woodward, Directors of Edgar Allen & Co., Ltd., and Messrs. S. J. Hewitt and L. F. Keeley, of the same firm, for help afforded in reading through and correcting the original and the revised MS., and other metallurgists and experts of the same firm for information, specimens, and photographs. Thanks are also due to my firm, Edgar Allen & Co., Ltd., of Sheffield, without whose co-operation and general sympathy and encouragement the publication of this book would have been much more difficult.

ERIC N. SIMONS.

SHEFFIELD, *1937*.

CONTENTS

CHAPTER I

Mixtures, Compounds, and Solutions

Iron and Carbon.

Steel is a crystalline substance. A definition given in a work of reference describes it as " that form of iron produced in a fluid condition, and hence practically free from slag (difference from wrought iron), which contains less than about 2·20 per cent of carbon—as a rule less than 1·50 per cent (difference from cast iron) ". Another accepted definition is: " Iron which is malleable at least in some one range of temperature, and in addition is either cast into an initially malleable mass, or is capable of hardening greatly by sudden cooling, or is both so cast, and so capable of hardening."

The most important constituent of steel is the element carbon. Carbon gives the strength and hardness to steel, but the increase in strength is at the expense of ductility and malleability. Further, in order to obtain this increased strength the carbon must not be present as carbon, but chemically combined with iron as a compound known as *carbide of iron*. If the carbon is present as " free " carbon, i.e. as *graphite*, the steel is useless for all practical purposes, except as scrap for remelting. With cast iron it is a different matter; " grey " cast iron contains most of its carbon as graphite, and, indeed, in this material free carbon, as graphite, is a necessary constituent.

Chemical Compounds: Cementite.

It is necessary for the reader to understand clearly what is meant by a *compound*. When two or more substances are mixed together they may or may not, according to circumstances, retain their separate individualities. If they do, then we describe the mixture as a *mechanical mixture*. Thus if we mix dry sand and dry sugar the result is a mechanical mixture, and no matter how thoroughly mixed, the separate particles of sand and sugar may be seen lying side by side and unchanged, if examined with a sufficiently powerful hand lens. Moreover, the colour of the mixture is intermediate between that of the yellow sand and that of the white sugar: the greater the proportion of sugar in the mixture, the less yellow is its colour.

When some substances are mixed, however, they combine chemically, often with the evolution of heat, and the particles of the original substances lose their separate individualities. They have, in fact, given rise to the formation of an entirely new substance or *chemical compound*, which possesses entirely different properties. Water is a chemical compound of hydrogen and oxygen, yet it is a liquid which does not burn or support burning and, as is well known, will extinguish a fire. On the other hand, hydrogen is an inflammable gas and oxygen is a gas which supports burning.

If we mix iron filings with powdered sulphur we form a mechanical mixture, and with a pocket lens we can distinguish the yellow sulphur particles and the bright metallic iron particles lying side by side unchanged. Now let us heat this mixture until it melts, say over a gas burner. The two elements now combine to form a chemical compound which is known as iron sulphide, or *sulphide of iron*, and when this is cooled its colour is seen to be almost black, and entirely different from that of the iron-sulphur mixture from which it has been formed. Examination with a pocket lens no longer reveals the separate iron and sulphur particles. Further, if we treat this compound with dilute spirits of salts (hydrochloric acid) a foul-smelling gas known as hydrogen sulphide is given off, and if a sufficient excess of the acid is used, the whole of the sulphide of iron dissolves. If the original iron-sulphur mixture is treated with the acid, however, only the iron is dissolved, the yellow sulphur remains untouched, and the gas which is given off is hydrogen, not hydrogen sulphide.

It will be evident, therefore, that the properties of a compound are altogether different from those of a simple mechanical mixture of its elements.

Similarly, when carbon is added to iron i: forms the compound we call *carbide of iron*, and this has entirely different properties from those of iron or carbon. The latter, either as charcoal or graphite, is a soft brittle substance which is easily powdered, while pure iron is a soft, malleable, and ductile metal. Carbide of iron, on the other hand, is an intensely hard substance, and although brittle as compared with iron, is not powdered so easily as carbon. Carbide of iron is often called *cementite*, and has the chemical formula Fe_3C.

Steel is really an alloy of iron and carbon or, more correctly, an alloy of iron and iron carbide. In annealed and normalized steel the carbide of iron can be observed under the microscope; the terms annealing and normalizing are defined later (pp. 53–54). In heat-treated steels, however, particularly hardened steel, the carbide particles can no longer be separately identified, even under a very powerful microscope.

Solution: Solid Solution: Austenite.

When steel is heated, at a certain temperature which depends on its composition, the carbide dissolves in the iron of which it is largely composed, to give what is described as a *solid solution* of carbide in iron.

The whole theory of the metallurgy of steel treatment is based on the formation of solutions, so that it is important to understand exactly what the term " solution " really means. Thus, consider the solution or dissolving of sugar in a cup of tea. The tea is sweetened and the sugar is evenly distributed throughout it, but in such a finely divided or dispersed form that not even the most powerful microscope would reveal its presence. *The sugar has not combined with the tea but has dissolved in it to give a solution of sugar in tea.* It is often said that sugar " melts " when it is added to tea. This is wrong, however, notwithstanding popular usage, and melting and dissolving are altogether different phenomena. To melt a material it must be heated, and generally it is heated alone.

When steel is heated to a certain temperature, the tiny particles of carbide of iron dissolve or go into solution in the iron, and above this temperature, although the steel is still quite solid, the particles of carbide could not be observed under the most powerful microscope even if it were possible to examine the hot steel under these conditions. The iron carbide and the iron have formed a solution, but since it has been formed when the steel was solid it is said to be a *solid solution*. This solid solution of carbide of iron in iron has a special name: it is known as *austenite*. It is entirely different in its properties from annealed or normalized steel (pp. 53–54). Annealed steel is a mechanical mixture of the hard carbide and the soft iron, and its properties are intermediate between those of the two substances of which it is composed. This explains why carbide increases the strength but lowers the ductility of annealed steel.

The properties of austenite are quite different from those of mixtures of iron and carbide of iron.

Saturation: Precipitation: Phase.

Now let us consider once again our solution of sugar in tea. We cannot go on putting lumps of sugar into a cup of tea, and expect every one to be dissolved. After a certain point the sugar ceases to dissolve and remains as an undissolved solid at the bottom of the cup. When this point is reached the solution of tea is said to be *saturated* with respect to the sugar, and the solution of tea and the undissolved sugar are said to be " in equilibrium " with each other. The saturation or equilibrium point depends on the temperature of the solution. One of the first things taught in chemistry is that heat increases the power of one substance to dissolve another; cold, on the other hand, decreases it. If, therefore, we heat a cold saturated solution of tea and sugar we shall be able to add a little more sugar and actually see it dissolve. If, on the other hand, we begin to freeze our saturated solution, some of the already dissolved sugar will cease to be retained by the solution, and will return to solid form and fall to the bottom of the cup. The solid sugar thus thrown out of solution is then said to be *precipitated*.

Now it is certainly not quite true that the saturation point for a solid solution of iron carbide in iron is reached in so simple a manner as this, but the principle involved is exactly the same. Precipitation of carbide of iron from saturated austenite may also be likened to that of precipitation of sugar from a saturated solution of sugar in tea.

The term *phase* is frequently employed in connexion with the structure of steel. Iron and steel when heated, or cooled from a very high temperature, pass through a series of different physical conditions. Thus, at temperatures above its melting-point (1535° C.) pure iron is completely fluid, and is then said to exist in the liquid phase. Below this temperature the iron exists in the solid phase, but, as we shall see later, the solid iron successively changes its physical form as it cools, and each of these different physical forms constitutes a different phase. Each different phase is physically and chemically uniform.

CHAPTER II

Equilibrium Diagrams

Freezing of Mixtures.

The points discussed in the previous chapter bring us at once to a consideration of *equilibrium diagrams*, without an understanding of which it is impossible to follow the changes in structure which take place during the cooling of steel.

Let us suppose that we have two different molten metals, gold and silver. If we pour one of these liquid metals into the other we obtain a perfectly uniform liquid solution. Now let us allow this liquid solution to cool. It can be shown by experiment that it begins to freeze at a set temperature (which depends on its composition), and continues to freeze until another set temperature is reached, below which it is completely solid. In other words there is a definite temperature range of freezing or solidification. Moreover, the extent of this range varies according to the proportion in which the gold and silver are present. In this series of alloys it is only when the composition is 100 per cent of gold or 100 per cent of silver that freezing takes place at one fixed and invariable temperature.

To make this clearer: the freezing or solidifying temperature of pure gold is 1063° C. and no other. But suppose we have a liquid solution containing 60 per cent of gold and 40 per cent of silver. On cooling it begins to freeze (solidify, or crystallize) at approximately 1050° C., and continues to freeze until the temperature has dropped to about 1000° C. There is thus a range of about 50° C. for the solidification of this alloy. Reduce the gold to 20 per cent and increase the silver to 80 per cent, and the freezing range is from about

1005° C. to 965° C., i.e. 40° C. Thus both the solidifying temperatures and the freezing ranges, as indicated, vary according to the proportions of the alloying elements.

Equilibrium Diagram: Eutectic: Eutectoid.

It is customary to show these ranges of freezing on a diagram in relation to the percentages of its components. Fig. 1 is a simple diagram of this kind. It shows the freezing or solidifying ranges for salt-water solutions. In this diagram the lines

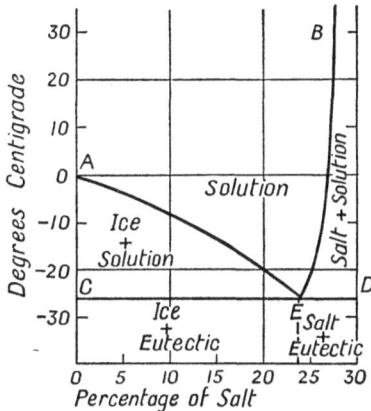

Fig. 1

AEB indicate the temperatures at which freezing *begins*, and the line CED the temperature at which any solution becomes completely solid. E is a particularly important point, and is known as the *eutectic point*. The line AE shows the temperatures at which solutions containing less salt than E begin to solidify, and BE the temperatures at which solutions containing more salt than E begin to freeze. In the former case, it is ice that first freezes out, and in the latter salt first freezes out. When the solution contains 23·5 per cent of salt, i.e. when the solution has eutectic composition, it freezes at a fixed temperature, and both ice and salt freeze out side by side. The eutectic solution has no freezing range; its freezing temperature is fixed and invariable. The word " eutectic " is derived from the Greek and actually means " easily melting ".

Let us now consider the freezing of different solutions in a little more detail.

When pure water is cooled its temperature falls regularly down to zero (0° C.), at which it begins to freeze. During the whole time of freezing, its temperature remains at 0° C. until finally every drop of water has been converted into solid ice. If a small percentage of salt, say 2 per cent, is added to the water, freezing does not begin at zero, but at some lower temperature, say —2° C. From this temperature down to —22° C. crystals of ice are deposited. This naturally means that the remaining liquid solution has become progressively richer in salt and, in fact, when the temperature reaches —22° C. the liquid solution remaining has the composition corresponding to E, i.e. it has eutectic composition. At this temperature the liquid freezes, and deposits small crystals of ice and salt together as an intimate mechanical mixture. This mechanical mixture of the two solid phases, ice and salt, is known as the *eutectic*, and its composition and freezing temperature are absolutely fixed and invariable.

If more salt is added to the original 2 per cent solution, the first formation of ice occurs at a still lower temperature than —2° C., but the temperature of final solidification remains the same, i.e. —22° C. The liquid which finally freezes at this temperature, of which there is proportionally more, contains again 23·5 per cent of salt. Thus the eutectic is invariable in its proportions whatever the composition of the original solution. If exactly 23·5 per cent of salt is present in the initial solution, there will be no ice formed until the eutectic point is reached, when it is deposited side by side with salt, so that the whole of the solution freezes at —22° C. If the salt in the original solution exceeds 23·5 per cent, then salt first begins to form instead of ice, but eventually at —22° C.

the final liquor contains exactly 23·5 per cent of salt and then freezes as the eutectic, as before. Thus, whatever the original percentage of salt, the final composition at the eutectic temperature is always the same.

These principles at once show why ice on footpaths and roads in winter can be melted by sprinkling with salt.

Now the principles outlined above for salt and water solutions apply equally well to the study of solutions of carbon in iron, i.e. to liquid iron and steel. It will be noticed that the freezing lines AE and EB form roughly a letter " V ". When a V-shaped curve of this type is shown on an equilibrium diagram, it can always be associated with the formation of either a eutectic or a eutectoid. The essential difference between these is that whereas the former is produced during actual freezing, i.e. during the period of final solidification, *eutectoids* are formed entirely from solid solutions instead of from liquid solutions.

Equilibrium diagrams are of great value to the metallurgist. They show at a glance the compositions and freezing-points of eutectics and other substances. The equilibrium diagram for iron and steel is much more intricate than the salt-water and gold-silver systems, but with the above explanations as a preliminary, we are in a better position to grasp the significance of the matter.

CHAPTER III

Freezing of Iron-Carbon Alloys

Iron-Carbon Equilibrium Diagram (first part).

We may now proceed to a consideration of the freezing of iron-carbon alloys, and a simple reproduction of part of the equilibrium diagram appears below (fig. 2). Molten steel consists, as has been said, of carbon (or carbide) dissolved in liquid iron. Before freezing can occur, the temperature must fall until a point is reached on either of the lines AB or BC. These two lines, it will be noticed, are similar to the lines AE and EB on the salt-water diagram (fig. 1, p. 7). We have this difference, however, between the two diagrams. In fig. 1 the end of the freezing range is represented by the horizontal line CED, but in the iron-carbon diagram the end of freezing or solidification is represented by the line AEBF. From this diagram it should be clear to the reader that the iron-carbon eutectic alloy is a mechanical mixture containing the equivalent of 4·3 per cent of carbon. When this eutectic freezes, however, its two components are the austenite solid solution, having the composition E, and carbide of iron (not carbon). An alloy containing exactly 4·3 per cent of carbon would consist entirely, when just solid, of the eutectic mechanical mixture. If the carbon content is less than 4·3 per cent but more than about 1·8 per cent, the austenite solid solution is first deposited on freezing, and more and more of the austenite is deposited down to about 1120° C. At this temperature the remaining liquor has eutectic composition and then freezes at a constant and *fixed* temperature.

With all alloys containing between 1·8 per cent and 4·3 per cent of carbon the liquid which finally freezes has the eutectic composition, so that conditions are similar to those which we found in our salt-water system. With alloys containing less than 1·8 per cent of carbon, however, the freezing takes place in a similar manner as with the gold-silver alloys, i.e. for *each* alloy there is a range of temperature over which freezing

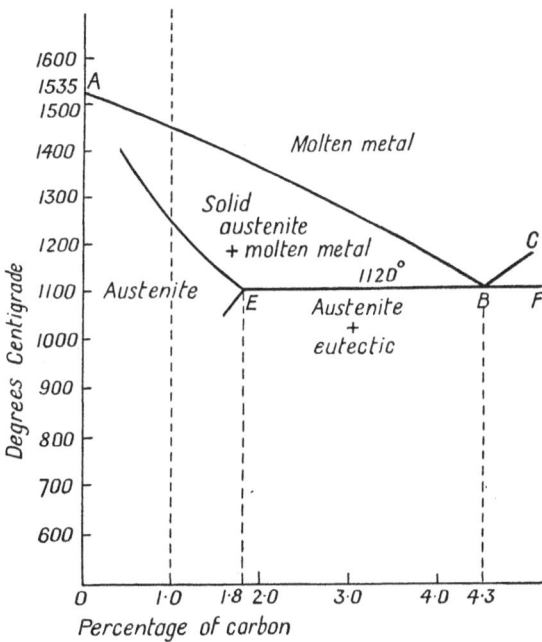

Fig. 2

takes place, but the final liquid to freeze does not reach eutectic composition, and the structure of the solid alloy consists entirely of the austenite solid solution. Theoretically, this austenite should be perfectly uniform and have the same composition as the original liquid. A steel containing about 1·0 per cent of carbon begins to freeze at about 1450° C. and is completely solid at about 1250° C.

This difference between alloys containing less and more

than 1·8 per cent of carbon is used by metallurgists for convenience in order to distinguish between cast iron and steel. Alloys containing less than 1·8 per cent of carbon are called steels, and those with more than 1·8 per cent of carbon are cast irons. It should be pointed out, as indeed has been indicated, that so far as the above reasoning is concerned, our conclusions relate to cast irons containing the carbon as carbide. These are known as " white " cast irons. The structures of the " grey " cast irons containing much free carbon in the form of graphite can be accounted for in a somewhat similar way, but since in this book we are not very much concerned with grey cast iron, it is not proposed to go any further into this matter beyond what has already been stated.

To revert to steel, we have seen that its structure when it has just solidified consists of the austenite solid solution, but we have no information yet as to what happens in the steel when it is further cooled down to room temperature. In the case of alloys of gold and silver, once they have solidified there are no further structural changes, but with steel there are several important structural changes which take place in the solid steel as it cools still further. The nature of these changes will now be explained.

Iron-Carbon Equilibrium Diagram (second part).

In order to follow what happens when iron and steel cool down to normal room temperatures it is necessary to make use of a further diagram, shown in fig. 3, which is often described as the steel part of the iron-carbon equilibrium diagram. In some respects it resembles the diagrams examined previously (there is a V-shaped precipitation curve, for instance), but there are important differences to which reference is made later. It has already been explained that austenite is the solid solution of carbide of iron in iron and not merely a mixture. Now let us examine the behaviour of this important substance on cooling down from temperatures below that of final solidification.

Eutectoid: Ferrite: Cementite: Pearlite.

Consider, for example, the point O in the diagram of fig. 3, representing the structure of a steel containing 0·5 per cent of carbon at a temperature of 1000° C. At this temperature the steel is composed of austenite crystals, as indicated, but on cooling, at a temperature of about 780° C. (point X on line AE) begins to reject or deposit *pure iron*. As the tempera-

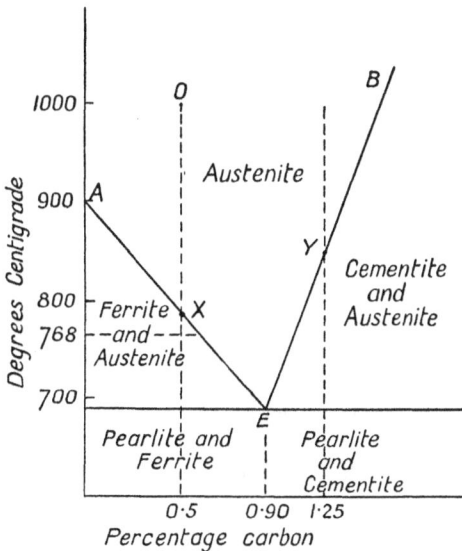

Fig. 3

ture continues to fall, more and more *pure iron* is deposited from the austenite, until eventually the austenite solid solution remaining contains the equivalent of 0·9 per cent of carbon at 695° C. At this temperature this remaining austenite deposits pure iron and carbide of iron side by side, just as previously salt and ice were deposited side by side from our salt solution of eutectic composition. Here, however, the two substances are being deposited from a *solid* solution instead of from a liquid solution, and in order to differentiate between the two, we use in the present case the term *eutectoid*. The eutectoid in steel therefore contains the equivalent of 0·9 per cent of carbon and is composed of pure iron and carbide of iron or, to be more precise, is a mechanical mixture of 13·5 per cent of carbide of iron (cementite) and 86·5 per cent of pure iron (ferrite).

If we consider now the cooling of a steel containing 1·25 per cent carbon, we find that when the point Y (approximately 855° C.) on the line EB is reached, the austenite of which it is composed is beginning to deposit carbide of iron (cementite), and this deposition continues down to 695° C. This means that proportionally more carbon than iron is being removed from the austenite, so that when 695° C. is reached, the remaining austenite, as before, has the eutectoid composition and then deposits iron and carbide of iron together.

We have so far learned the meanings of the terms austenite, eutectic, and eutectoid, and their meanings should be borne in mind in all that follows, as they will be referred to again and again in reference to the metallurgy of steel. Two other common terms have also just been mentioned, viz. *ferrite* and *cementite*. It has been seen that on cooling a 0·5 per cent carbon steel, practically pure iron is precipitated from 780° C. down to 695° C. The iron thus precipitated is known as *ferrite* to distinguish it from the iron of the eutectoid. On the other hand, it has been pointed out that in a 1·25 per cent carbon steel carbide of iron is first deposited at about 855° C. and deposition continues down to 695° C. This precipitated carbide of iron is known as *cementite* in order to distinguish it from the carbide in the eutectoid and, moreover, the term cementite is applied to any carbide in steel, even when it contains alloys such as manganese, chromium, and the like.

Another term may conveniently be explained here. We have seen that the " eutectoid " in steel is a mechanical mixture formed at 695° C. having an *average* carbon content of 0·9 per cent of carbon, and consisting of cementite and ferrite in

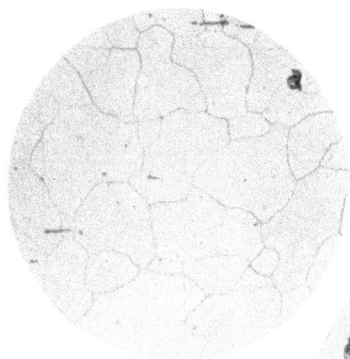

Fig. 4

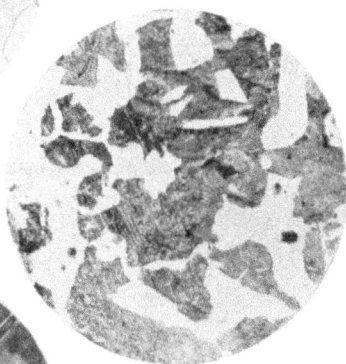

Fig. 5

Fig. 6

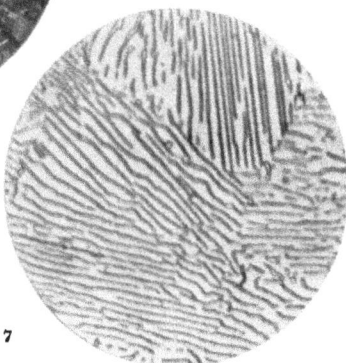

Fig. 7

Facing p. 14

definite proportions. This eutectoid mixture is described as *pearlite* by metallurgists.

We are now obtaining an insight into the structure of steel. Here we may point out that so far as plain carbon steels are concerned, the temperature of 695° C. represents the point of final structural change, or at least so far as we need go in this book.

Photomicrographs.

It is important, having traced the changes which take place as steel cools from the molten condition, that we should now begin to understand the *appearance* presented by the various constituents to which reference has been made. To obtain pictures of these constituents and the structural condition of any steel or metal, metallurgists prepare specimens in a special way. Pieces of the steel to be examined under the microscope are given a plane surface rendered perfectly free from all scratches by cutting, filing, and polishing with the proper abrasives. Certain constituents are revealed by polishing alone owing to differences in colour or hardness—chiefly in non-ferrous alloys—but usually the specimen must be treated with chemical reagents in order to bring up the differences in colour or appearance between the various constituents. This process is called *etching*. There are other means of obtaining metal structures but these need not yet be considered.

Pearlite consists, as we have seen, of cementite and ferrite. The illustrations given (figs. 4–7) are what are known as *photomicrographs*, which are photographs taken by a special method when an etched specimen is magnified under the microscope. The first of these photomicrographs shows ferrite; the second shows ferrite and pearlite; the third shows cementite and pearlite; whilst the fourth, which is at a much higher magnification, shows the structure of pearlite itself. The carbide of iron and the ferrite of the pearlite are clearly revealed in the form of alternate plates. It is this laminated structure of pearlite which is responsible for the " mother-of-pearl " appearance when it is etched, hence the name " pearlite ".

We are now in a position to consider the structure of steel in greater detail and from a more technical standpoint, and also to visualize the complete equilibrium diagram. This will be done in the next chapter.

CHAPTER IV

Critical Points

Steel and White Cast Iron: Complete Diagram.

In the preceding chapter it was pointed out that when steel cools down after solidification, structural changes occur, resulting in the formation of a solid structure consisting of either ferrite and pearlite; cementite and pearlite; or pearlite alone. The mass will consist of ferrite and pearlite whenever there is less than 0·9 per cent of carbon in the steel; of cementite and pearlite, when there is more than 0·9 per cent of carbon; and of pearlite alone, when there is exactly 0·9 per cent of carbon in the steel. The cementite and ferrite are respectively the excess carbide of iron and pure iron precipitated out in order that the eutectoid, " pearlite ", (with 0·9 pe cent carbon) may be formed. We have seen what these structures look like.

Up to now, we have considered only what happens to steel when cooled down from the point of solidification. But we have not yet examined the white cast irons.

It was pointed out in Chapter III that *white* cast irons are those containing more than 1·8 per cent of carbon (as carbide) and that their structures just after complete solidification consist of austenite plus the eutectic of austenite and carbide. When cooled further the austenite puts down more carbide (cementite) until finally the austenite itself (at 695° C.) has eutectoid composition and then changes to pearlite. The essential difference between steel and white cast iron is, therefore, in the amount of free carbide (cementite) present in the structure of the cast iron at normal temperatures. In

17

white cast iron the amount of cementite may be greatly in excess of the pearlite, and it is this which accounts for its brittleness and fragility. Although this book is concerned not so much with cast iron as with steel, we mention the structure of cast iron because many of the high-percentage alloy steels, e.g. high-speed steel, high-carbon steel, high-chromium

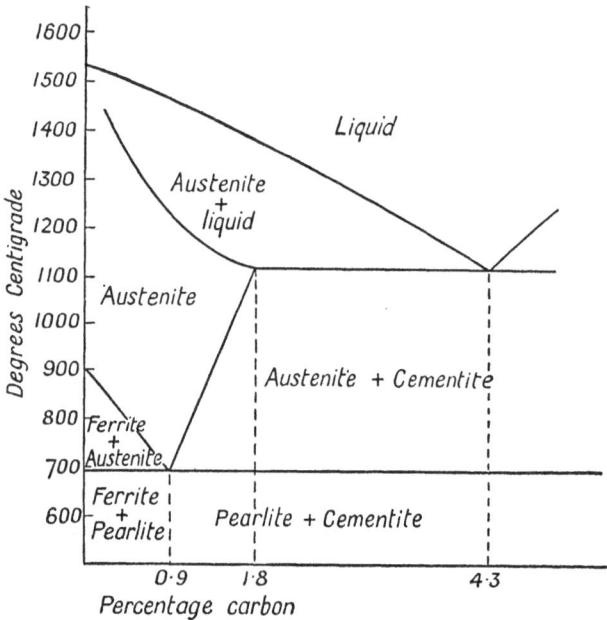

Fig. 8

steels, &c., have similar structures to that of white cast iron. The complete iron-carbon diagram which deals with steel and white cast iron is now shown in fig. 8.

There are still other points which need to be considered before we can proceed much further with the structure of steel. It has been seen that in cooling steel from, say, 1000° C. down to 695° C., the austenite is transformed into either (a) ferrite + pearlite, (b) cementite + pearlite, or (c) if the carbon content is exactly 0·9 per cent, into pearlite alone. These structural changes are accompanied by other physical changes, which must be understood before progress can be made.

Cooling Curves.

Suppose now that we have at our disposal the necessary apparatus for measuring accurately the temperature of steel as it cools from a temperature above its freezing point and that we measure the temperature at regular intervals of time. By this means we can find out whether the fall in temperature is regular or otherwise. We should find, in actual fact, that during the actual freezing or solidifying period the rate of cooling was less than when the steel is either completely liquid or completely solid, i.e. there is a lag or slowing-up, or delay in the cooling during the period of solidification. Since we have not in any way altered the environment of the steel (e.g. brought it into contact with something hot or held it in front of a fire), such a retardation of the cooling can only be due to a giving out of heat by the steel itself, i.e. the molten steel gives out heat as it solidifies. Heat evolution takes place in a similar way when solid steel cools. Thus if we cool a steel containing 0·9 per cent of carbon from 1000° C., its temperature falls regularly down to 695° C. At this temperature, however, it remains stationary for a period depending on the size or mass of the steel. It should be evident that this " halt " or " arrest " in the cooling has some connexion with the formation of pearlite from austenite containing 0·9 per cent carbon.

Time-Temperature Curves.

Since an understanding of cooling curves is a matter of great importance, let us go back to very simple examples such as pure water and solutions of salt in water. Suppose we start with lukewarm water and cool it, noting its temperature after regular time intervals. If we plot the results, i.e. note them down on a chart and connect them by lines to show their movement graphically, we obtain the curve shown in fig. 9 (i). This is called a *time-temperature* curve. It will be seen that the temperature falls regularly until 0° C. is reached, when it remains stationary until the water has been changed to ice, after which the temperature continues to fall regularly. Now let us start with a warm solution of 10 per cent of salt in water (compare fig. 1, p. 7) and repeat our cooling experiments.

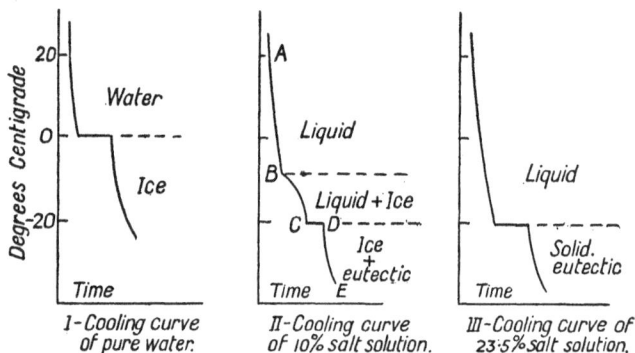

I—Cooling curve of pure water.   II—Cooling curve of 10% salt solution.   III—Cooling curve of 23·5% salt solution.

Fig. 9

The resulting time-temperature cooling curve is shown in fig. 9 (ii). This may be divided into four different parts: A–B, where the liquid solution is cooling regularly; B–C, where the saturated solution is depositing or precipitating crystals of ice and the rate of cooling is retarded; C–D, the horizontal part which corresponds with the freezing of the eutectic; and D–E, where the temperature of the completely solid mass is cooling regularly. If we started with a solution containing 23·5 per cent of salt, the cooling curve would be as shown in fig. 9 (iii), and it will be seen that the only difference between this curve and that for pure water is the matter of the temperature of the horizontal portion of the curve.

Inverse-Rate Curves.

When a eutectoid such as pearlite in steel is formed, a similar horizontal branch is shown in the cooling curve. When dealing with changes in solid alloys, however, the temperatures associated with the structural changes are often not very clearly defined by means of time-temperature curves, and much more reliable information is then obtained by plotting in a somewhat different way, i.e. we ascertain the number of seconds taken for the temperature to fall 1° C. When the results are plotted we get what is called an *inverse-rate curve*, and this shows up the temperatures associated with the structural changes in a more clear and accurate manner. The inverse-rate curves for water, 10 per cent and 23·5 per cent salt solu-

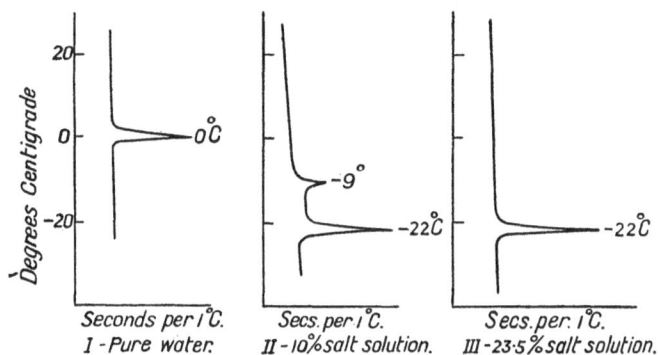

Fig. 10

tions, are shown in fig. 10, and these should be compared with the time-temperature curves shown in fig. 9, when it will be seen that the important temperatures are now indicated by " peaks ". The way is now paved for an examination, in the next paragraphs, of the relation between cooling curves and the equilibrium diagram.

Arrest-Points: Allotropic Forms of Iron.

In the preceding section examples of cooling curves were given, and it was pointed out that the most important temperatures are shown on inverse-rate cooling curves by one or more points or *peaks*. These peaks are described as *arrest-*

points, or *critical points*. Now if we could obtain the inverse-rate cooling curve for pure iron from the perfectly molten condition down to room temperatures, our curve would be like that shown in fig. 11. The first critical point occurs at 1535° C., and marks the freezing-point of pure iron; the second occurs at 1404° C.; the third at 900° C.; and the

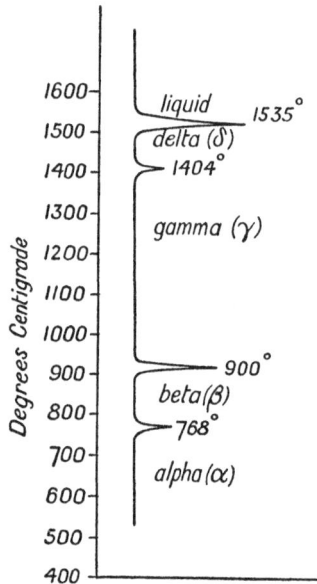

Fig. 11

fourth and last at 768° C. The question at once arises: what is the significance or meaning of the arrest-points which occur in solid iron? By referring back to fig. 3 (p. 13) it will be seen that the 900° critical point coincides with the point A and thus has something to do with the formation of ferrite. But here we are dealing with pure iron alone, so that all these critical points must be connected with changes which take place *within the iron itself*, as it cools. Actually, this means that iron is really a different substance physically within different ranges of temperature. When an element is capable of existing in more than one form, it is termed *allotropic* and the various forms are known as *allotropic modifications* of the element. One of the most familiar elements which exhibits allotropy, apart from iron, is carbon, which exists in the form of diamond, charcoal or soot, and graphite.

Between its freezing-point and 1404° C., iron is known as *delta* (δ) iron; between 1404° and 900° C. as *gamma* (γ) iron; between 900° and 768° C. as *beta* (β) iron; and below 768° C. as *alpha* (α) iron. Thus the iron at ordinary temperatures which we have described as ferrite is also α-iron.

The physical and mechanical properties of these allotropic forms possess important differences. Thus whereas α-iron is a strongly magnetic substance, both β- and γ-iron are non-magnetic and are not attracted by a magnet. *Actually, however, this is the only difference between α- and β-iron—in all other respects they are alike—so that we do not now regard them as separate and distinct allotropic forms. β-iron is simply non-magnetic α-iron and its formation has no influence on the properties of heat-treated iron or steel.* Having mentioned β-iron we shall not worry much about it in the rest of the book, but simply regard it as non-magnetic α-iron. We were compelled to mention the existence of β-iron, of course, particularly on account of the fact that in some of the older and out-of-date textbooks several peculiar properties are wrongly ascribed to it.

Modern research has shown also that δ-iron is essentially the same as α-iron, so that in reality there are only two different allotropic forms of iron, viz. alpha and gamma. The main difference between these so far as we are concerned is the fact that γ-iron is the only form capable of dissolving carbon to any appreciable extent. The solubility of carbon in α-iron is very small in comparison, and it is upon this difference in carbon solubility that the whole of the heat-treatment of steel is based.

The solid solutions of carbon (as carbide of iron) in γ-iron are called *austenite*. Before pearlite (in which the iron is present as ferrite) can be changed to austenite, the ferrite in it must be transformed to γ-iron.

Notation for Arrest-Points.

In addition to naming the various forms of iron, it is convenient, for reasons which will be evident later, to represent the critical points in iron and steel by letters and numbers. Thus the 1404° C. change-point (δ-γ) is known as the Ar_4 point. The 900° C. arrest, i.e. the γ-β change, is described as Ar_3, and the magnetic change at 768° C., i.e. the so-called β-α point, as the Ar_2 point. By these representations they will be described in the pages which follow.

The relation of these critical points to the equilibrium diagram still remains to be considered. As indicated, the 900° C. point coincides with the point A in the diagram on p. 13 (fig. 3), the magnetic α-β change being indicated by the dotted horizontal line at 768° C., from which it will be gathered that the temperature of the magnetic change is uninfluenced by carbon. The 1404° C. (δ-γ) change-point was deliberately not indicated in the equilibrium diagram, because it might have introduced unnecessary complications. This change is of some importance in connexion with the structures of low carbon steel ingots, but does not usually concern us in the heat-treatment of steel.

Hitherto in this discussion we have considered pure iron; it now remains to consider the effects of the introduction of carbon. As soon as carbon is introduced we get the appearance of another critical point at 695° C., which increases in prominence with increasing carbon content up to 0·9 per cent. This new critical point is described as the Ar_1 point, and it should now be evident that it is associated with the formation of pearlite from austenite.

Another effect of the introduction of carbon into iron is the lowering of the Ar_3(γ-α) critical point as, indeed, is indicated by the downward slope of the line AE in fig. 3 (p. 13). When the carbon content of the steel equals 0·6 per cent the Ar_3

point is coincident with Ar_2. This combined point is then progressively lowered by further additions of carbon until with 0·9 per cent carbon there is one arrest-point only. Thus a 0·9 per cent carbon steel, on cooling, has one critical point at 695° C.

We have now considered the changes which occur during cooling, and shall next turn our attention to the phenomena which occur during the heating of steel.

It has been shown that when iron and steel are cooled, a giving out of heat occurs at certain temperatures where the amount of heat evolved is sufficient to retard the normal rate of cooling. It is only reasonable, therefore, to assume that if we reverse the process, and heat steel from the cold state, the reverse phenomenon will occur, i.e. there will be an absorption of heat at certain temperatures sufficient to retard the normal rate of heating. This reasoning is shown, in practice, to be perfectly correct. Just as there are critical points on cooling, so there are corresponding critical points on heating, and these are named in a similar way. It might be imagined that as we are completely reversing the process, these critical·points would coincide exactly with those obtained during cooling and that the same names would do for both. But this is not so, and the critical points on heating are always higher than the corresponding cooling (Ar) points. New designation numbers are therefore necessary, so that we have the Ac_1, Ac_2 and Ac_3 heating points corresponding to the Ar_1, Ar_2 and Ar_3 points on cooling. The Ac_2 (the α–β magnetic change) is about 770° C. as compared with 768° C. for Ar_2, i.e. there is very little difference. The Ac_1 and Ac_3 points, however, are always about 30° C. or so higher than the corresponding Ar points. Typical inverse-rate heating and cooling curves are shown in fig. 12.

The difference between corresponding Ac and Ar points is described as *thermal lag*, or *thermal hysteresis*, and the extent of this lag is an important point in connexion with heat-treatment operations. In general, the amount of lag is in-

creased by the addition of alloying elements.   As fig. 12
shows, the Ar_1 point, i.e. the austenite-pearlite point, is about
695° C., but the Ac_1 point representing the reverse change,
i.e. pearlite to austenite, occurs at about 730° C.

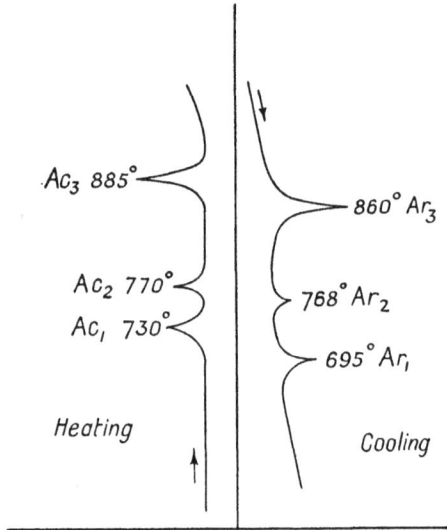

Fig. 12.—Heating and cooling curves of 0·20% carbon steel

Physical Properties.

We may now summarize the properties of the allotropic
forms of iron in regard to their physical behaviour. *Alpha*
(or α) iron is soft and ductile, and is present in the steels of
the softest and most ductile character, and also in wrought
iron.   Its tensile strength is about 20 tons per square inch.
Beta (or β) iron, as already indicated, is merely non-magnetic
α-iron and, apart from this difference, has similar properties.
At one time β-iron was supposed to be hard, but this view was
disproved by metallurgists a long time ago. *Gamma* (or γ)
iron and its austenite solid solutions are also soft and plastic,
softer even than α-iron (ferrite).   It is chiefly on this account

that steel is heated to the austenite region so that it may readily be forged and rolled to shape.

One final point must be mentioned in this chapter. Carbon, as pointed out in previous sections, is the most powerful element in its effect on the properties of steel, and gives, in proportion to the increase in its content, greater strength and hardness than proportionate increases of other elements. While this is mainly due to the influence of the iron carbide which is formed, and which is present in annealed and normalized steels, the hardnesses of steels in either of these conditions are not really so very extraordinary. An annealed steel containing 0·9 per cent of carbon has a tensile strength of about 37 tons per square inch, as against that of 20 tons per square inch for pure iron. When a steel containing 0·9 per cent of carbon is heated to a temperature above the Ac_1 point, and rapidly cooled by quenching in water, a truly phenomenal increase in hardness results, and the question at once arises: to what is this increase in hardness due, in view of what we have already said in regard to the hardness of ferrite (α-iron) and austenite (γ-iron solid solutions)? The fact is that a new constituent, known as *martensite*, is formed when steel is quenched from temperatures above its upper critical point. The properties and probable constitution of martensite are discussed in later sections.

Hitherto we have discussed what changes take place when iron and steel are cooled down from the molten state, but how these changes occur has not been discussed. To consider in detail how these metals solidify, it is necessary to examine the phenomena of crystallization, and this will be done in the following chapter.

CHAPTER V

Crystalline Forms

Atoms: Molecules: Crystals.

In what follows an attempt will be made to explain, as simply as possible, the actual way in which iron and steel become solid. We have discussed in a general manner the phenomena accompanying their solidification, but the reader will need to go further than this in order to obtain a more complete understanding of the structure of steel.

Although many readers will be familiar with the general outline of molecular physics, whereby the atom is regarded as a conglomerate of electrons, positrons and neutrons, we shall, in these articles, regard the atom as our unit in dealing with steel and its structure.

We have used the term *element* and must now explain it. In the first place, it may be stated that pure iron is an element and steel is not. The reason is that in a piece of iron all the atoms are iron atoms and are exactly alike, whereas in a piece of steel there are carbon, manganese, sulphur, phosphorus and perhaps other atoms in addition to iron atoms. An element is a substance whose atoms are all alike and possess the same chemical properties. Steel, therefore, is not an element. Atoms, even of elements, are not invariably detached and separate. They possess affinities for each other and for atoms of other elements, and sometimes exist in pairs. Oxygen, for instance, is a gas made up of molecules, each consisting of two oxygen atoms.

Occasionally, the normal conditions in which a molecule of a substance exists are greatly changed. This may result in a regrouping or redistribution of the atoms. For example, extreme heat may break up or dissociate, as it is called, a chemical compound into its separate elements. At very high temperatures water vapour is dissociated into its elements, hydrogen and oxygen. Ordinary oxygen is composed of molecules each containing two atoms. When three atoms of oxygen are held together, the substance produced is ozone, which is usually supposed to be present in greater amounts in the atmosphere at the seaside. Some elements and compounds are continually striving to revert to a more stable atomic arrangement or configuration, whereby they discard or gather atoms. Such substances are termed *unstable*; ozone, for instance, is actually an unstable substance because, if given the slightest opportunity to do so, it discards its third oxygen atom, and reverts to the more stable two-atom structure of ordinary oxygen. A most important point to bear in mind is that changes in atomic or molecular structure are accompanied by changes in physical and chemical properties.

We now come to a series of facts of great importance. Atoms may be gaseous, liquid or solid, and when an element is converted from one form into another, it is still composed of atoms. Thus the atom of iron in solid iron is identical with the atom of iron in molten iron, and if we could obtain gaseous iron, it would still consist of iron atoms. When the state of an element is changed, what is done is to strengthen or weaken the cohesion of the atoms or molecules, but not to alter their chemical character.

When solidification occurs, however, the atoms or molecules are not found situated at random, like the currants in a cake. They group themselves, curiously enough, into regular patterns or, rather, shapes. These shapes are of geometrical character, and are known as crystals. A large batch of crystals united with each other constitutes a mass of the given substance.

Our knowledge of the true structure of crystals is comparatively recent, but it is now established that they consist of a beautiful and symmetrical arrangement of their atoms. This brings us to the most up-to-date and satisfactory definition of a crystalline mass, i.e. one characterized by a regular arrangement of its structural units.

Multitudes of crystalline forms exist. Many of these can be seen on our window panes in snowy weather. Most important for us, however, are the crystals of iron and steel, and these and their characteristics will now be discussed.

Crystals of Iron and Steel.

The crystals or " grains " of which iron and steel are composed are built up of exceedingly small cubes, although the way in which the atoms are arranged in these cubes is different in the several allotropic modifications of iron. The atomic crystal unit cells are of two types, represented in figs. 13 and 14. The black dots represent atoms, but it should be mentioned that the lines joining them are purely imaginary. The

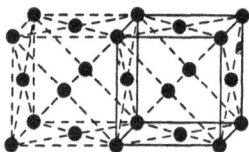

Fig. 13.—" Face-centred " cubic lattice of γ-iron

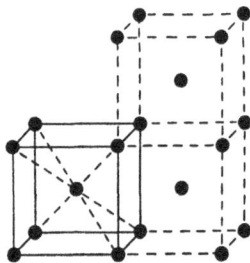

Fig. 14.—" Body-centred " cubic lattice of α- and δ-iron

arrangement or spacing of the atoms in this way is described as the *space-lattice*, i.e. it indicates the framework upon which the atoms are situated. The first diagram (fig. 13) is known as a *face-centred cubic lattice*, because in addition to an atom at each corner, there is also an atom at the centre of each face of the cube.

In fig. 14 there is an atom at each corner, as before, and also one atom in the middle of each cube; an arrangement called a *body-centred cubic lattice*. Gamma iron is face-centred and alpha iron is body-centred cubic. These cubes are exceedingly small; the length of each side of the cube is less than a tenth of a millionth of an inch, so that there are literally millions of them to a cubic inch of iron or steel. Actually, when alpha iron changes to gamma iron the length of each side of the cube increases, but this expansion is more than counter-balanced by the larger number of atoms in each cube, so that the net result is actually a contraction. Similarly, when gamma iron changes to alpha there is a net expansion.

The small cubic crystal cells cannot, of course, be seen even under the most powerful microscope. The methods by which the above results have been deduced are explained in Chapter XIII (p. 107). It must be realized, however, that iron and

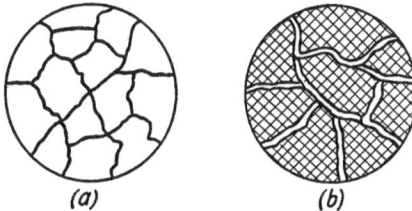

(a)   (b)

Fig. 15

steel are built up in this way and are therefore truly crystalline substances, although they may not appear to be so, when seen by the eye or when examined by the microscope. What we do actually see under the microscope with pure iron or any other pure metal is something like that shown in fig. 15 (a). Each of these irregularly-shaped figures is called a *crystal* or *crystal grain*, and each is built up of an exceedingly large number of the atomic cubical units. In each separate crystal the axes of the small cubes all point the same way, but in different directions from one crystal to another, an effect known as *orientation*. It is this fact which makes them different crystals; if all the axes of the cubes in a piece of iron pointed in the same direction, it would be one large crystal.

Cleavage Planes: Crystal Boundaries.

Now let us picture in our minds the thousands of cubic cells in a single grain. A little imagination only should indicate that the atoms are situated, as it were, in several sets of parallel planes, known as *cleavage planes*. It is like taking a child's wooden brick and slicing it horizontally and vertically so as to form smaller cubes, then repeating this process indefinitely. The slicing planes correspond to cleavage planes. Cleavage planes, whenever they exist, are a real cause of weakness. They are present not only in iron and other pure metals but in other substances such as rocks, many of which, as is well known, split quite readily along certain planes. Similarly, when iron is broken, the fracture or splitting is along the cleavage planes of its crystals.

At the junctions or boundaries of different crystals there are, of necessity, spare atoms which do not occupy regular or symmetrical positions on the space-lattice, i.e. they are positioned at random. There are not enough atoms to " go round ", as it were, in each crystal, so that in the actual boundaries there are no cleavage planes, and the boundaries do not split so readily, i.e. they are stronger than the crystals themselves. This explains why fracture most readily takes place along cleavage planes, i.e. *through* the crystals and not along their boundaries. Further, the smaller the crystal size, the larger the number of boundaries that must be broken across before fracture can take place. A small or fine-grained structure is thus always stronger than a larger- or coarse-grained structure of the same metal or alloy.

The above remarks in regard to the greater strength of crystal boundaries only apply to pure metals and uniform solid solutions. When impurities or other constituents are present it may be an entirely different story. During cooling, a pure metal throws out the impurities or other foreign atoms. These may then form layers or films separating the crystals of the parent metal from each other. Fig. 15 (*b*) shows, very crudely, this phenomenon. The cross-hatched parts represent the pure metal or solid solution and the white streaks represent the rejected substance or impurity.

Now consider a cast plain carbon steel containing, say, 1·3

per cent of carbon. As previously explained, it consists of the eutectoid, pearlite, and free cementite. The pearlite, or rather the austenite from which it was formed, is equivalent to a pure metal, so that the superfluous or free cementite is rejected to the grain boundaries. As previously indicated, however, cementite is hard but very brittle, so that in a 1·3 per cent carbon steel in the cast condition, the structure consists of relatively ductile pearlite grains separated by the excess brittle cementite. Thus the structure as a whole is relatively weak and easily broken or fractured. We thus arrive at this very important conclusion: *the strength of a metal or alloy is largely a matter of the strength (or fragility) of its crystal boundaries.* With a pure metal or uniform solid solution the boundaries are stronger than the crystals themselves, and fracture takes place across the crystals along the cleavage planes. When, however, the boundaries contain impurities or other brittle constituents, the fracture may occur along the crystal boundaries or separating surfaces. And so, by easy stages, we are brought to a consideration of grain size, which will now be dealt with.

Grain Size.

When we take a piece of steel and break it, the resulting broken surfaces are termed " fractures ", and the study of these fractures (even with the naked eye alone) tells us a great deal about the quality and character of the metal. But with the fracture polished and etched, the microscope tells us much more. The individual crystals or grains can be detected, and their size is a very good index of quality. If the grains are large and coarse, the steel will probably be inferior, or in an inferior condition. If small and close, a higher quality of steel is suggested. The reasons for this have just been explained.

We cannot, in this book, go deeply into the highly scientific and elaborate research work which has been devoted to grain size and its significance. We will content ourselves with choosing from the vast bulk of observations and experiments certain facts which bear upon our subject.

The first point which has been proved beyond question both practically and by metallurgical research is that smaller grain size is always associated with increased toughness and strength.

The next important discovery bears largely upon our forthcoming study of hardening, tempering and annealing as applied to steel. It is this: when we begin to cool a steel, the final grain size is affected by the temperature at which the steel stood when cooling began. The higher the temperature above the critical point known as Ac_3 when we begin our cooling operation, the larger and coarser will be the grains or crystals in our final product. If, on the other hand, we begin our cooling operations from the Ac_3 point (i.e. if in our first operations we do not heat much above the Ac_3 point), then our final product will have the best possible grain size. In fact, just as there is a definite relation between toughness and grain size, so there is a relation between grain size and the temperature from which cooling begins.

We now come to the third point. However irregular, coarse, and large the grains of a steel may be, the moment we heat them (i.e. the steel they compose) to a temperature just above the Ac_3 or other upper critical point, the large crystals disappear and are replaced by the best possible grain size.

Research has taught us yet more about grain size and its effect. It is now known that to heat a steel for any considerable period above 1000° C. greatly enlarges the crystals and causes embrittlement. Hence the importance of care in heating steel. Sustained overheating does irreparable harm.

There is one point relative to steel and iron crystals that needs mention before we pass on to a closer examination of steel structure as revealed by the microscope. Fortunately for civilization and the steel manufacturer, the cleavage planes of iron and steel crystals do not all lie in the same direction, or almost any bar could be cracked by a comparatively light blow, whatever the size of its grains. It is because in commercial irons and steels the crystals lie at all angles to each other that these metals possess the strength that is their greatest recommendation. We must remember, too, that although fineness of grain is usually synonymous with good quality, it can so happen that fineness of grain is a bad rather than a good quality. Into this, at the moment, however, we need not go.

Cold steel, then, under the microscope, is seen to consist of crystals or grains of pearlite interleaved with, or surrounded by, crystals of ferrite or, alternatively, of pearlite grains and cementite crystals, together with whatever additional elements or impurities the steel contains. *Solid steel, in short, is always crystalline.*

Annealing and normalizing are two technical terms which, to some extent, define the method of cooling. With annealed steels, these are allowed to cool down slowly in the furnace, after being soaked at the proper temperature, whereas normalized steels are withdrawn from the furnace and allowed to cool down in the air. Generally the structure of the normalized steel is finer, but it must be realized that the two terms cannot be put into " water-tight " compartments; the cooling of a large mass or piece in air may yield an annealed structure.

Martensite: Troostite: Sorbite.

In a preceding chapter we described certain of the constituents of steel as seen under the microscope, e.g. ferrite, cementite, pearlite and austenite. There is no need to go over this ground again. There are, however, other constituents of steel with which we have not yet dealt, and the present seems a suitable opportunity for discussing them, in order that when we come to other phenomena of steel in which they play a part, we may refer to them by name without the necessity for explanation.

The first of these hitherto ignored constituents is martensite, which was first discovered by A. Martens, a metallurgist, and director of the royal mechano-technical experimental station at Charlottenburg, Germany. He published a paper on the " Microscopic Investigation of Iron " in 1878, and went a long way in correlating the application of the microscope to the study of fractures and visual characteristics.

Martensite is really a primary stage in the breaking down of austenite, which, as we have seen, is an unstable substance excepting at high temperatures. Austenite is rarely found in ordinary steels and generally changes readily on cooling into its constituents, ferrite and cementite. Martensite is a transitional substance arising out of this breakdown of the austenite. When a steel containing, say, 0·9 per cent of carbon is heated above its critical point and then rapidly cooled by quenching in water, it has been shown that the critical point on cooling occurs at about 300° C. At this relatively low temperature the steel is so rigid that it cannot yield to the impressed volume change which accompanies the change from gamma to alpha iron, and its structure is therefore that of a strained alpha lattice with, perhaps, the minute distribution of carbide along the cleavage planes. The exact nature of martensite is even to-day not completely understood, but the above is a generally accepted view of its constitution.

Martensite is hard, harder, in fact, than any other constituent of steel containing an equal amount of carbon. It is the essential constituent of any hardened tool steel.

To harden steel and obtain the hard martensitic state it is necessary to cool it faster than a rate known as the *critical hardening speed* which, for carbon steels, is very rapid indeed. With alloy steels the critical hardening speed is much less, so that in some cases cooling in oil or, in certain cases, simply cooling in air suffices to yield the hard martensite. Martensite is strongly magnetic.

If a steel is rapidly cooled but at a rate less than the critical hardening velocity, another constituent, known as *troostite*, makes its appearance. This constituent is named after the French metallographist, L. Troost, and is another of the transitional substances arising out of the breakdown of austenite into pearlite, but is formed at a much higher temperature than martensite. Troostite is formed when carbon steels are quenched in oil or when large sections are quenched in water. The hardness of troostite is about half that of martensite.

Sorbite, named after that grand old Sheffield scientist, Dr. H. C. Sorby, is a third of these marginal or transitional constituents, and may be found in normalized carbon steels relatively high in manganese or in the interiors of large masses quenched in water. When thus formed it has a great resemblance to pearlite; so great, in fact, that it is not always easy to distinguish them; but its characteristics are different, and for this reason it must be kept in mind. It is stronger by far than pearlite, and its structure is much more intricate and close. The true sorbite is produced by first hardening the steel and then tempering it at about 650° C. The exceedingly fine sorbitic structure thus formed possesses a unique combination of tenacity and ductility.

These various constituents are, at first sight, extremely confusing. It will simplify the matter if we show, in a kind of genealogical table, the place of each in the scheme of things, as below:

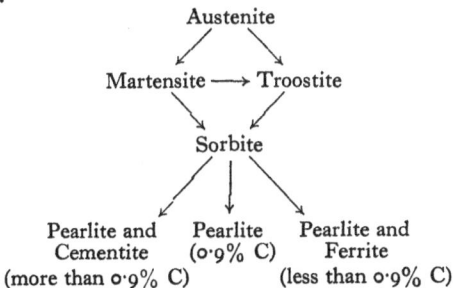

Austenite

Martensite ⟶ Troostite

Sorbite

Pearlite and Cementite (more than 0·9% C)   Pearlite (0·9% C)   Pearlite and Ferrite (less than 0·9% C)

Hitherto we have dealt with iron and carbon only. We must now discuss the effect of impurities and alloys on the structure and properties of steel. This will be attempted in the next chapter.

CHAPTER VI

Impurities of Steel

Impurities in Iron and in Steel.

It must now be clearly indicated that commercial iron and steel are never completely pure, and that the impurities they contain inevitably affect their grain structure and properties. Let us now consider some of these impurities and their effects. The chief impurities found in iron, apart from carbon, are silicon, sulphur, phosphorus, and manganese. These are solids. It almost invariably happens, however, that gaseous impurities also occur, e.g. nitrogen, oxygen, and hydrogen. In steel the same impurities are found, but it must be remembered that, although they are impurities in the real sense, they can (or at least some of them, such as silicon and manganese) also become alloying elements with beneficial effects, as will be seen later.

For the moment, however, we shall regard them as essentially impurities in the mild and medium carbon steels with which we are dealing.

Silicon.

The first to be considered is *silicon*, which is found in small quantities in every type of steel, usually from 0·10 to 0·20 per cent, but ranging from as low as 0·05 per cent to as high, on occasion, as 0·5 per cent or even higher in certain special alloy steels. Silicon in steel is present in solid solution in the iron, and slightly increases the strength and hardness, also raising the critical points.

Sulphur.

A much more harmful element existing in steel as an impurity is *sulphur*. Anything over 0·1 per cent in steel is usually regarded as making it practically unusable, and the lower the sulphur percentage the better the quality of the steel. Efficient steel manufacturers often contrive to reduce the sulphur to as low as 0·015 per cent or even, on occasion, to as low as 0·010 per cent. Sulphur has an affinity for manganese which, as indicated above, always occurs in steel. This is an important fact, because if it were not for this affinity for manganese, sulphur would combine chemically with the iron to form iron sulphide (FeS), which, for reasons shortly to be stated, is an even less desirable compound in steel than manganese sulphide. Consequently, the steel manufacturer has to take care that there is enough manganese in the steel to absorb the whole of the sulphur percentage. In practice, this means that anything from twice to eight times as much manganese as sulphur is necessary for this purpose, and in general eight times is better and more certain than any smaller proportion.

If sulphur is allowed, owing to insufficiency of manganese, to form iron sulphide, trouble is certain to occur, for iron sulphide, besides being far more brittle than manganese sulphide, does not occur in the structure of the cold metal in the form of solid particles or globules, but exists in the form of thin exceedingly brittle films separating the grains of ferrite and pearlite. This has a disastrous influence on the strength of the steel, as will be realized from a consideration of previous remarks on crystal formation. In the absence of sufficient manganese, steel is rendered both " hot-short " and " cold-short " by the presence of sulphur, i.e. it tends to break down and disintegrate when worked, either hot or cold.

Phosphorus.

Another impurity, almost as harmful as sulphur, is *phosphorus*, and, moreover, its harmful influence is less easily combated, although it has recently been shown that the bad effects of phosphorus are much less marked when the steel contains a little copper and/or chromium. Phosphorus readily combines with iron, and forms a phosphide of iron having the chemical formula Fe_3P. In high phosphorus pig irons and cast irons it may exist in the form of this phosphide, but in steel the phosphorus is usually in solid solution in the iron, and then increases the hardness and tensile strength, but, unfortunately, it also seriously affects the ductility and resistance to impact or shock. In addition, for a given treatment, phosphorus tends to increase the grain size of the steel. Its influence is, therefore, to be regarded as definitely bad, and although as much as 0·05 per cent is allowable in structural steels, the phosphorus content of a good tool steel should in no circumstances be greater than 0·035 per cent.

One of the greatest troubles associated with sulphur and phosphorus is that known as *segregation*, i.e. concentration of these two elements during freezing in areas between the austenite crystals of which the steel is composed. In these areas the sulphur and phosphorus values may greatly exceed the average percentages. Segregation is often found in the top parts of ingots and castings.

In *free-cutting* mild steels both sulphur and phosphorus are intentionally present in amounts up to 0·15 per cent each. Such steels are very readily machined, but a high manganese percentage is necessary in order to reduce the bad influence of the sulphur to a minimum.

Manganese.

Our next impurity is *manganese*. This element is also an essential constituent of any steel. It is a powerful and most effective deoxidant and, apart from its influence on sulphur as stated above, is perhaps the most useful element for the production of perfectly sound steel free from internal defects such as *blowholes* or gas-cavities. Manganese also improves the tenacity of steel without seriously affecting its ductility. Moreover, it increases the " depth-hardness " of hardened steel. Steels low in manganese may have a very shallow hardened skin even when quenched in water. When very high percentages of manganese are added, say between 10 and 18 per cent, and the carbon content of the steel is about 1 per cent, the steel is retained in the austenitic condition when cooled in air or sprayed with water after being heated to about 1000° C. The steel is then non-magnetic, and although soft, as measured by the Brinell test, it possesses a remarkable resistance to abrasion. Whether present as an impurity or as an intentional constituent, therefore, manganese has a beneficial influence on the properties of steel.

Gaseous Impurities.

Of the gaseous impurities *oxygen* is perhaps the most important. In general, it has a bad influence on the properties of steel. It rarely exceeds 0·025 per cent, but the percentage is greatest in Bessemer steels and least in steels made in the electric furnace; Siemens open-hearth steels have intermediate values.

Nitrogen is found in percentages ranging from 0·005 to as much as 0·3 per cent in welds. Nitrogen has a hardening and embrittling tendency. Here, too, the amount is greatest in Bessemer steel and least in electric furnace steel. Nitrogen is intentionally introduced into the surface layers of certain case-hardening steels, resulting in the production of an intensely hard skin which will withstand abrasion. In these *nitrided steels* nitrogen thus becomes an alloying element rather than an impurity. In welds it is an undesirable constituent.

The effect of *hydrogen* in steel is bad; its presence in steel is responsible for the formation of gas cavities near the skin of ingots and castings, and thus leads to unsoundness. The formation of " flakes " has also been attributed to hydrogen. It may be asked: where does the hydrogen come from?

There is little doubt that its presence is due to water-vapour in the furnace charge or atmosphere.

We now come to the fascinating and complex subject of alloy steels. This will be considered, in part at least, in the next chapter.

CHAPTER VII

Alloy Steels

Carbon Steels and Alloy Steels.

In the preceding chapter, certain solid or gaseous inclusions in steel were described as impurities. This is because their occurrence is accidental, unavoidable, and usually harmful. Certain additions of alien substances are made to steel intentionally, however, either to produce certain specific results, or to improve the steel generally. Such additions are known as alloys, and the steels containing them are known as alloy steels, in contradistinction to the carbon steels, in which none of these extraneous materials occur. Carbon steels are themselves, in actual fact, " alloy " steels, but since every commercial steel contains some carbon, the term " alloy steels " is, in common parlance, confined to steels containing one or more added elements.

The enormous development in numerous branches of the engineering, automobile, chemical and other industries is due, almost entirely, to the growth and multiplication of alloy steels, which have been the main feature of steel progress during the present century. It would be impossible, in this little book, to give a comprehensive account of the multitudinous alloy steels and their uses, nor, from the point of view of steel structure, is it essential. All that will be attempted is a rapid survey of the principal alloys and their effect on the structure of steel.

Silicon Steels.

Since *silicon* was chosen as the first example of an impurity in steel, it will be interesting to regard it now in the light of an alloy, as in silicon steels.

As remarked upon in the last chapter, all steels contain some small proportion of silicon, but there are certain alloy steels in which considerably larger amounts of this element are found. Thus the *silicon-steel* (really silicon-iron), invented by Sir Robert Hadfield, contains about 3·5 per cent of silicon, and very little carbon. The structure of this steel consists of silico-ferrite, i.e. the whole of the silicon is in solid solution in the iron. Silicon-steel is magnetic but possesses a low magnetic permeability, i.e. it is easily magnetized and demagnetized, and on this account is widely used in transformers, where it is necessary to cut down magnetic losses to a minimum. Then there are the *silico-manganese steels* used for springs, containing about 0·55 per cent C, 1·25–2·00 per cent Si, and 0·8–0·9 per cent Mn, and the *silico-chrome steels* used for exhaust valves, containing about 0·40 per cent C, 3·5 per cent Si, 8·0–9·0 per cent Cr.

It was mentioned in Chapter VI that silicon raises the critical points in steel. So does chromium, and the critical point in silico-chrome steel is relatively very high, a useful practical feature since it means that there is less likelihood of successive hardenings and temperings during heating and cooling, when the valves are in use. The critical points, and therefore the hardening temperatures, of silico-manganese spring steels are also considerably higher than those of steels low in silicon but of otherwise similar composition. One detrimental influence of silicon in steel is the tendency to promote decomposition of the carbide of iron into iron and graphite. If this occurs, the steel is generally useless. This difficulty is overcome by having enough manganese (and sometimes chromium) present in the steel. In grey pig iron and grey cast iron the silicon may be as high as 3·0 per cent, and it is this which largely accounts for most of the carbon then existing as graphite.

Manganese.

Manganese is extensively used as an alloying element in steel. The effect of this metal on the structure of steel is considerable. The critical points in steel are lowered by manganese, and when about 2·0 per cent is added, the critical points in steels of even moderate carbon content merge into one another, so that only one change-point occurs. When more than 2·0 per cent is added, the steels possess air-hardening properties, and their structures are martensitic at normal temperatures after cooling in air.

Nickel.

Nickel is a most important alloying element. Its main influence is to lower all the critical points in steel. With about 4·0 per cent of nickel the A_3 and A_2 points are merged into one another, and if still higher proportions of nickel are added, only one change-point may be obtained. The lowering of the Ar_1 point is greater than the lowering of the Ac_1 point, i.e. nickel increases the thermal lag or hysteresis, which was referred to in Chapter IV. Thus the Ac_1 point in a steel containing 0·40 per cent C and 3·75 per cent Ni was 710° C., and its Ar_1 point occurred at 595° C. Nickel also refines the grain size of steel, but to obtain the best results for structural purposes from pearlitic nickel steels, which can be heat-treated just as carbon steels, we must lower the carbon content as the nickel percentage is increased.

If the nickel is increased up to more than 30 per cent, the critical point in steel is below zero, so that on cooling its structure consists of the non-magnetic austenite.

Chromium.

Another most important alloying element is the metal *chromium*, which is the key-element in the wide range of stainless and heat-resisting steels developed in recent years. Chromium readily dissolves in liquid iron, and is also held in solid solution in both alpha and gamma iron. When the chromium exceeds about 12·0 per cent, the alloys of iron and chromium no longer exhibit allotropy, but consist of ferrite solid solutions at all temperatures from atmospheric up to their melting-points.

In the presence of carbon some, at least, of the chromium is present in the carbide which separates from the austenite on cooling. In actual fact, the carbide or cementite in chromium steels is a solid solution of chromium carbide in iron carbide. This " complex " carbide dissolves much more sluggishly than pure iron carbide when the steel is heated above its critical point.

Chromium also alters the eutectoid carbon content, e.g. when the chromium content is about 13·0 per cent the amount of carbon in the eutectoid, pearlite, is about 0·3 per cent only, instead of 0·9 per cent. Chromium also raises the critical points in steel.

It must be borne in mind that alloys to-day are not only added singly. Chromium-nickel, nickel-chromium-molybdenum, cobalt-chromium-tungsten, are but three of the numerous alloy combinations in steel quite frequent in modern metallurgy. When these extensive alloy combinations, together with the widely varying percentages of the respective alloys, the fluctuations in carbon content, and the fact that different characteristics are exhibited at different temperatures, are borne in mind, it will be realized at once that no brief account like the present can effectively and intelligibly deal with the structural variations, the permutations and combinations, that occur. For an account of these, the reader must be referred to the specialized textbooks and treatises that abound.

Vanadium.

To revert to the simple alloy additions, one may mention next *vanadium*. This goes into solid solution with both alpha and gamma iron. When the amount of vanadium exceeds about 1·10 per cent, the iron-vanadium alloys do not undergo allotropic changes and consist of ferrite solid solutions at all temperatures, just like the iron-chromium alloys containing more than 12·0 per cent of chromium. In steel containing carbon, vanadium carbide is formed, which also enters into solid solution with carbide of iron. This complex carbide is even more sluggish in going into solution in gamma iron than the complex carbide in chromium steel. This is an advantage, since it means that vanadium steels are not so readily overheated as other steels.

Molybdenum.

Molybdenum is a further alloying element of commercial interest. Like chromium and vanadium, it goes into solid solution in both alpha and gamma iron, and forms complex carbides in the presence of carbon. Molybdenum appears to refine the grain size of steel, and is also introduced into nickel-chromium steels to avoid " temper-brittleness ", i.e. low resistance to impact or shock after hardening and tempering, the latter operation being conducted at 600° to 650° C., followed by slow cooling.

Tungsten.

A mention of alloying elements in steel would be incomplete without reference to *tungsten*, so greatly used in the manufacture of high-speed steels. In some respects the effect of tungsten is similar to that of molybdenum. Thus, it enters into solid solution in both alpha and gamma iron, and yields a carbide of tungsten (WC) when carbon is present. Even as little as 1·0 per cent of tungsten has a remarkable influence on the grain size of tool steels containing about 1·0 per cent of carbon. When hardened, steels of this type have a much finer grain and fracture than steels of similar carbon content free from tungsten. Tungsten is an essential constituent of high-speed steel, upon which it confers the property of " red-hardness ", i.e. the ability to cut even though the nose of the tool may be at a dull red heat. At such temperatures the cutting-edge of an ordinary carbon steel would be completely destroyed. Tungsten also raises the critical points in steel; hence the need for hardening high-speed steel at temperatures between 1250° and 1300° C.

Cobalt is another element introduced into certain special

steels. It is added to magnet steels to improve their " remanence " or, in other words, their stability under varying
magnetic conditions, and, in addition, is added to high-speed
steels to improve " red-hardness ", i.e. cutting efficiency at
elevated temperatures.

The manifold structural intricacies of these innumerable
alloy steels can be imagined, and their behaviour on heating
and cooling are of more interest to the metallurgist and research
worker than to the practical man. We therefore leave them at
this point, and go on to consider a series of operations of
enormous practical importance, comprised in the generic
term " heat-treatment ".

CHAPTER VIII

Heat-Treatment : Case-Hardening

Heat-Treatment and Structure.

Steel in the condition in which it is usually supplied is comparatively soft. For many purposes it has to be made harder, and to achieve this, a process of heat-treatment is necessary. This heat-treatment or " hardening " process is of great practical importance, and in order that it shall be carried out successfully, some knowledge of the metallographic and structural aspects of the matter are desirable. We will, therefore, review it first in the light of explanations already given.

In earlier chapters the relation between equilibrium diagrams and cooling curves has been studied, and it has been shown also that while steel changes its structure from austenite to pearlite when slowly cooled, other constituents may be formed when the rate of cooling is quickened. All these points are of great significance in the heat-treatment of steel, and must be thoroughly grasped in order that heat-treatment may be carried out satisfactorily.

It has been seen that if we heat a simple steel containing 0·9 per cent of carbon to a bright red, its structure will be austenitic. Cooled in the furnace or air, when the temperature reaches 695° C., the austenite changes to pearlite (ferrite and cementite). If quenched in water from above this temperature, the normal separation of the ferrite and cementite from the austenite is prevented. Nevertheless, little or no austenite is found in the quenched structure. Instead, if the cross-section of the steel is not too large, its structure is composed of martensite throughout. The probable constitution of martensite was discussed in Chapter V, where it was suggested that the hardness of martensite is due essentially to the straining of the alpha iron lattice in consequence of the carbide trying, as it were, to form but being unable to do so.

With these few comments in mind the reader will perhaps be able to follow with greater facility the actual phenomena of hardening, tempering, and annealing.

Hardening.

To harden a steel properly is a scientific operation demanding an accurate knowledge of the hardening or quenching temperature. The observed effects of sudden cooling of steels from temperatures between their critical points reveal that the higher the temperature of the steel before quenching, the greater is the hardness of the suddenly cooled steel, the ductility becoming less, until the upper change point is passed, after which the temperature of quenching has very little influence on the hardness but may seriously affect its toughness, i.e. render the steel more brittle. If, for instance, a 0·9 per cent carbon steel is heated to a temperature well above its critical point (730° C.) and then quenched, it is really overheated, and although it is certainly hard, is far too brittle for commercial use. A golden rule is never to heat any steel for hardening to a temperature much above its upper (Ac) critical point.

But there is another factor which has an important effect on the properties of the steel after quenching. This is the speed with which the piece is cooled. If this varies, the hardness of the steel will vary. It can be said, simply, that the quicker the cooling, the harder the steel. Certain quenching media are much more drastic in their cooling action than others. Iced or brine water cools hot steel more rapidly than water at room temperature, and water at room temperature more rapidly than whale oil. We choose our quenching medium, therefore, according to the degree of hardness it is desired to produce.

The effect, structurally explained, is that the quicker the cooling, the lower the temperature at which structural changes in the steel occur, and the more promptly these changes are arrested. The carbon is prevented from precipitating out, and the necessary hardness is obtained.

Yet another factor must be borne in mind in steel hardening, namely, the influence of mass or size. The fact that small pieces can be more thoroughly hardened than large ones is definite and not inexplicable.

Bullens gives the following data to represent the effect of mass on the tensile strength of steel:

Diameter of bar (inches) ..	$\frac{1}{2}$	1	$1\frac{1}{2}$	2	$2\frac{1}{2}$	3	$3\frac{1}{2}$
Max-stress (tons per sq. inch) ..	63	60	58	55	50	48	45

The reason is that as the piece increases in size, the longer is the time required for the effect of the quenching to pass from the surface to the interior. The heat is therefore extracted more slowly from the interior and its rate of cooling is retarded.

Recalescence.

It has been shown that when we cool a piece of steel containing 0·9 per cent of carbon, it shows a critical point at 695° C. Now if we could observe this cooling in a darkened room, it would be seen that the piece would actually glow at the temperature of the critical point, owing to the heat energy given out as the ferrite and cementite are deposited from the austenite. This phenomenon, i.e. the sudden evolution of heat at the Ar point during cooling, is often described as *recalescence*.

Tempering.

We have already dealt with hardening and must now consider the operation of *tempering*. Hardened steel may be very brittle and before it can be placed into service must be tempered, i.e. reheated to some temperature which is always below that of its lower critical point. To put the point broadly, there are two ranges of tempering: (a) low-temperature tempering, as applied to high-carbon tool steels, &c., where most of the initial hardness is retained, and (b) high-temperature tempering, up to 650° C., where any martensite is completely destroyed and replaced by the tough sorbite. The latter treatment is applied to the lower carbon structural steels, which are required to be tough and strong, yet capable of withstanding sudden shock and fatigue.

Low-temperature tempering of hardened steel may not lower the hardness much but will, nevertheless, most certainly lessen its brittleness by equalizing the internal strains developed during the hardening.

As the temperature of tempering rises, the hardness falls off but the toughness and ductility are improved. One essential feature of the tempering treatment is that it can be completely controlled so as to yield the necessary combination of hardness, strength and toughness. This is rendered possible solely because the martensite obtained by quenching is really an unstable constituent. It should not be present at all, and when the steel is reheated, it seizes the opportunity to revert to the normal structure. On tempering, therefore, the martensite (and troostite) gradually suffers breakdown with the deposition of carbide of iron particles, which increase in size as the temperature is raised. At first these particles are sub-microscopic, but coalescence or mergence takes place until, after tempering at temperatures above about 550° C., the carbide is rendered visible under the microscope. Troostite also breaks down in a similar way. Eventually, therefore, by tempering at a sufficiently high temperature after hardening, the normal constituents of steel, ferrite and cementite, are reproduced but are not arranged in quite the same fashion. Instead of laminated pearlite, with or without free ferrite or cementite, the structure of hardened and "fully" tempered steel consists of small spherical particles of cementite uniformly distributed throughout the ferrite matrix. This constitutes a *true sorbite*, and it is this fine uniform structure of sorbite which is responsible for the splendid combination of tenacity and ductility obtained from a hardened and tempered steel.

Annealing.

Another important heat-treatment process is *annealing*, and there is often some confusion in the minds of steel users as to the difference between this process and tempering. The reader should study pp. 51 and 90 in conjunction with this section and the following.

Tempering always implies a reheating, but at temperatures below the lower critical point, and usually after hardening or quenching, or after cold work. On the other hand, in annealing, the temperatures employed are generally at least above the lower critical point, although on occasion a *sub-critical annealing* may be conducted, where the annealing range is immediately below the *lower* change-point. The difference becomes clearer if, for purposes of simplification, we substitute the word " softening " for " annealing ". Users often require a steel in the soft state, perhaps because they wish to machine it readily. Annealing is the process which achieves this. But it does other things as well. It frees the metal from such internal stresses as may have been set up by manipulation under heat (rolling, forging, casting, pressing, &c.). But in no sense has annealing any relation to the hardening and tempering processes, from which it differs completely. Annealing is essentially a hardness-removing operation.

Commercial annealing temperatures are to a large extent dependent on the carbon content. With carbon less than 0·9 per cent, the steel is heated just above its *upper* critical point, but if the carbon content is greater than 0·9 per cent, it is heated to a point just inside the critical range, in such a manner that the austenitic state is never wholly achieved. The usual annealing temperatures for the high carbon tool steels are between 760° and 780° C.

The required annealing temperature is maintained for a

definite period, and then the steel is permitted to cool down slowly to normal temperature. There must be no hit or miss about the operation. Both temperature and time must be accurately known, so that the desired result may be brought about. In the end, the steel is no longer hard, and in a position of arrested structural change (as on tempering). The structural changes of annealing will be dealt with later, where necessary.

Normalizing.

The next heat-treatment process of interest is *normalizing*, sometimes termed " grain refining " (pp. 89 and 91).

This, like annealing, is a cooling process. The difference between the two lies in the method. In annealing, the steel is heated up, and allowed to cool in the original furnace, whereas in normalizing, it is permitted to cool in air. Actually normalizing is not a very exact term, owing to the differences in the speed of cooling inevitable when the size of the piece varies. A thin section will quickly cool throughout its entire structure. A large mass will cool much more slowly, and therefore the resulting structure in the interior of the mass will be more like that of annealed steel.

The normalizing treatment has definite advantages. It gives to ordinary steel a much greater degree of tensile strength than will annealing, and the yield point and reduction of area are likewise improved. (These terms will be explained in a later chapter.) Furthermore, normalized steels frequently provide a superior surface when machined to that which identical annealed steels would give. Normalizing is particularly advantageous for bars that have been rolled and emerge from the rolls at a comparatively low temperature, since it improves their ductility, a measure of which is lost in the rolling, and produces a greater uniformity.

Case-hardening.

We now come to *case-hardening* (other names for which are *case-carburizing* and *case-carbonizing*). In many industries machinery is employed whose parts are subjected to both wear and repeated shocks. To meet these requirements a steel must be obtained which is hard enough to withstand wear, yet tough enough not to be readily fractured by shock, and at the same time low enough in first cost to be economical. Such a steel is with difficulty procurable at a low price, and for this reason a method of combining these properties in an ordinary carbon steel (or in certain alloy steels) has been invented and is known as case-hardening.

Case-hardening is based on the assumption that carbon makes a steel harder. If we want a hard surface to withstand wear, we can, it is assumed, obtain it if we can manage by some means to increase the carbon percentage of the skin or surface layer, while leaving the rest of the material inside in its original condition, which is tough and ductile. The actual process of introducing the carbon into the skin is called *cementation*.

The method is to take the part after it has been machined from an ordinary low-carbon steel, embed it in some substance of very high carbon content, and then heat it to a temperature above 900° C. If this heating is correctly carried out, the steel will slowly soak up carbon from the surrounding mixture, which is one of the properties steel possesses. The iron above 900° C. is in the gamma condition. It is possible to obtain almost any required depth of " case " (i.e. of hardened skin or layer) by regulating the temperature and the period during which the heating is maintained. Usually the layer is between 1 and 2 mm.

The best case-hardening temperatures lie between 900° and 930° C. and this latter figure should not be exceeded. The reason is that otherwise there is a serious risk of over-carburization of the case with the formation of a network of free cementite.

The aim in case-carburizing should be to produce a " case " of eutectoid composition, the interior or " core " being retained in an unchanged condition so far as its carbon content is concerned.

The reason for using low or medium carbon steels is (1) their cheapness and easy machinability, and (2) their readiness to absorb additional carbon.

Case-hardened steels need to be heat-treated after the cementation process is finished. It will be clear from what has been written in earlier sections that by diffusing additional carbon into the low carbon steel layer or case, we are actually bringing the steel at these points into a condition closer and closer to the eutectoid. We shall therefore have to harden and temper it if we wish to obtain maximum hardness of case combined with toughness of core.

In addition to overheating, case-hardening can be spoiled by undue prolongation of the heating period, which means that the carbon goes deeper and deeper into the steel, until a eutectoid mass throughout is obtained, and the part ruined.

It will be as well to study the facts of carbon diffusion more closely. Carbon by itself does not diffuse into steel. So true is this, that if carbon in a completely pure form were packed round a case-hardening steel and heated to the correct temperatures, scarcely any would diffuse into the skin or surface layer. It is now known that the carbon is carried into the steel by gases freed during cementation.

It is because pure carbon is of so little use in case-hardening that such substances as wood charcoal are never used alone for the purpose, but are always compounded with some other substance, such as barium carbonate. As the choice of case-hardening compounds belongs to the practice of steel treating rather than to the structure of steel, we shall pursue this aspect of the matter no further.

Nitriding.

One form of case-hardening which has come very much to the fore in recent years is *nitriding*, in which nitrogen gas is employed instead of a compound to give the hard outer surface required. Nitrogen itself does not affect steel to any great extent, but if a steel part or piece is subjected to heat in the presence of an ammonia atmosphere, a case-hardening effect of great value is produced. Ammonia contains nitrogen, of course, and this nitrogen unites with the iron in the part. The result is the formation of needles of iron nitride (Fe_2N), which is very hard, giving a coarse martensitic appearance to the structure. These nitride needles are precipitated from ferrite solid solutions, and give to it the extreme hardness perceptible. The nearer one comes to the core, the smaller is the proportion of nitrides to ferrite, until eventually the normal condition of ferrite-pearlite is attained, or retained.

It is important to note that aluminium (about 1 per cent) is apparently a necessary constituent of steel intended for nitriding, and it is believed that there is a combination of aluminium and nitrogen, forming aluminium nitride.

Another important point is that steels to be nitrided must possess a rather higher percentage of carbon than ordinary case-hardening steels, for the reason that some carbon is forfeited as a result of the process, and must be compensated for unless the core is to sacrifice some of its necessary strength. It is interesting to note that alloy steels are more suitable for nitriding than ordinary steels. The latter give a surface of great hardness but excessively brittle, and liable, therefore, to flake off in use. Steels with vanadium, chromium, titanium, and the like, in their composition give a less brittle case of greater depth.

Nitriding is usually carried out between 450° and 500° C., and the economic advantage of the process is that it requires no heat-treatment of the parts afterwards. The temperatures are well below any critical point, so that there is never any enlargement of the crystals or grains. Owing to the simplicity of the method, there is practically no warping or distortion. The temperature of 500° C. is, however, a maximum which must be observed or the results will be less satisfactory. Heat-treatment of the parts before nitriding is essential.

CHAPTER IX

Characteristics of Alloy Steels

Silicon Steel.

It may now be of value to deal a little more in detail with the normal characteristics of a few of the better-known alloy steels. These were discussed from a purely theoretical point of view in an earlier chapter (Chapter VI, p. 38). We will now deal with them more generally. Silicon steel, discovered in 1888 by Sir Robert Hadfield, has a number of interesting characteristics. Silicon raises the critical points at which structural transformations occur, which implies that silicon steels must be heated to a higher temperature for heat-treatment than the standard carbon steels. The steel containing it has a good degree of magnetic permeability and a high electrical resistance. Moreover, it is tough and stands up well to fatigue, even at raised temperatures. On the other hand, unless it is carefully and properly heat-treated, the grains tend to become coarse, which is a drawback.

Silicon has also the ability to break up carbide of iron. Silicon steels are principally used for transformers, in the form of laminated cores. In combination with manganese, silicon makes up a silico-manganese spring steel of commercial importance. In the transformer cores the silicon percentage is approximately $3\frac{1}{2}$, and in the spring steel, 1·25–2. Silicon steel should not be cold worked as this has a bad effect on the magnetic properties.

In tool steel, silicon does no damage so long as the percentage does not exceed 0·4 per cent, but any increase in this percentage is definitely harmful, though in certain special circumstances it may be higher. Silicon up to about 3 per cent is sometimes found in chromium valve steels for high temperature use, and up to 0·5 per cent in high chromium stainless steels.

Manganese Steel.

Manganese steel (12–14 per cent manganese) is the most important of the commercial steels containing manganese. It has a number of remarkable characteristics. The steel is normally in the austenitic state. A quenching treatment from 1000° C. results in the total retention of this austenite, with the iron in the gamma condition. This heat-treatment would leave an ordinary carbon steel hard and brittle, but with manganese steel it has precisely the contrary effect, leaving it as soft as is possible. (The Brinell number is as low as 200.) It is possible to bend it in the cold state, and a blow of sufficient force will make an impression without great difficulty. Yet the moment you try to cut it with a chisel or file it with a file, you fail. We will see why this is, shortly. Until the super high speed steels containing high percentages of cobalt (such as Edgar Allen " Stag Major ") were invented, it was commercially impracticable to machine this steel, and any finishing work on it had to be done by grinding. Now, however, given proper plant, conditions, and technique, manganese steel can be and is regularly drilled and turned, as a practical job.

Manganese steel can be forged. It is used in the form of castings and forgings (jaws, toggle-plates, liners, end-plates, &c., for crushing and grinding machinery; points, crossings, switches and rails for railways and tramways; pins and bushes for dredgers; &c. See booklet " Imperial Manganese Steel "). A quenching treatment is necessary to put the steel into the best state for the finishing operations.

Annealing makes the steel brittle. Tempering it in the quenched condition results in a change to the martensitic state. Martensite is much harder than austenite and there is a consequent loss of ductility. Manganese steel in the normal condition, being austenitic, is non-magnetic. The iron is in the gamma form, and only when some of this has been transformed into alpha iron does magnetism arise. Thus, as martensite consists largely of alpha iron, the tempering treatment makes the steel magnetic.

Now the martensitic transformation takes time to occur when brought about by heating, but there is another way in which this same martensitic transformation can be brought about. This is by cold working the steel. Thus, the moment a piece of cold austenitic manganese steel is abraded, filed, chiselled, or drilled, its surface instantaneously becomes martensitic and therefore too hard to be worked with any facility. Hence the astonishing resistance to wear offered by this material.

Manganese steel is never annealed, and if wrongly heat-treated, will become brittle and virtually incapable of being machined.

From the above explanation, it will be seen why the new high-speed steels have made the machining of the material possible. They do not simply rub the surface, making it harder and harder, and therefore less possible to cut. They cut straight through the surface before it has had time to work-harden and become martensitic. Hence the importance of keeping tool edges sharp, and not allowing them to become blunt so that they rub and lead to their own breakdown.

Cobalt Steels.

Cobalt steels have been referred to in the preceding paragraphs. These steels have two principal commercial uses, as high-speed cutting steels, and as steels for permanent magnets. In regard to the latter use, cobalt in percentages ranging from 3 to 35 has the effect of improving the magnetic properties of magnet steels. It increases the remanence to some extent, but its principal effect is to raise the coercive force, giving a much higher $B\text{-}H_{max}$ figure. The " remanence " (B_{rem}) is the amount of remanent magnetism left in a piece of steel after it has been subjected to a magnetizing force; the coercive force (H_c) is the amount of negative magnetizing force required to demagnetize a magnet, i.e. the magnetic force needed to remove the remanent magnetism. Cobalt steel is, however, expensive, hard, and difficult to forge. For this reason, in the cutting steels it is usually alloyed with chromium.

A few general notes on the structural effect of introducing alloys into steel may be advantageous at this point. Usually, in an alloy steel, the pearlitic structure needs less carbon for its formation than is the case with ordinary carbon steels. Secondly, alloys have the effect of changing the temperatures at which structural transformations occur, which means that alloy steels need different heat-treatments as compared with carbon steels. Lastly, the alloy additions delay the structural changes.

Chromium Steels.

Chromium steels are used in widely varying industries. The percentage of chromium ranges from 0·25 to 30, according to the purpose for which the steel is used. Typical uses are razors, cutlery, ball bearings; hacksaw blades; chisels; permanent magnets; rolls for cold rolling and for crushing stone, &c.; wearing parts of ball, tube, and rod mills; and dies.

Chromium combines very freely with iron, but is usually found in the commercial steels as a chromium carbide. It can, of course, go into solid solution in alpha or gamma iron. The chromium does not entirely drive out the iron from the iron carbide, but takes up enough carbon to form chromium carbide, and then, with the remaining iron carbide, forms a double carbide solid solution of great stability and hardness.

Chromium, in short, stabilizes the carbide in steel, and so eliminates the formation and release of carbon during the manipulatory and annealing processes. Up to one per cent raises the tensile strength of a carbon steel without reducing ductility to any marked degree. It increases the ability to withstand wear, and particularly that form of wear involving an abrasive action as distinct from a hammering action, in which respect it is superior even to manganese steel. Even in low percentages, chromium gives improved resistance to corrosion.

Both in cooling and in heating, the effect produced by alloying steel with chromium is to raise the change-points to higher temperatures. If, however, cooling is speeded up, the Ar (cooling-change) point is lowered. This means that chromium steels, to some extent, are self-hardening. One disadvantage of chromium is that it increases the brittleness of steel by reason of the fact that when the steel is heated, the grain size may be considerably increased. If the steel is accidentally overheated, the steel becomes excessively brittle. Hence, simple chromium steels have to be very carefully and scientifically heat-treated. That is why for so many purposes an additional element, such as nickel or vanadium, is added to reduce this danger of brittleness due to excessive grain growth.

Nickel Steels.

Nickel steels are among the most useful of all the alloy steels, but they are not quite simple alloys of steel and nickel alone. There is always a percentage of manganese in their composition. The reason is that nickel is exactly the opposite of chromium in respect to carbon liberation. It tends to *promote* the freeing of carbon from the carbide in the form of graphite. Some element must be added to stabilize the carbide, and usually about 0·6–0·9 per cent of manganese or chromium is chosen.

Nickel has a great effect on the change-points in steel. Fig. 16 shows this effect on a 3½ per cent nickel steel. The percentages usually added range from 2½ to 5, but there are one or two special corrosion-resisting requirements for which as high a percentage as 35 is used. These, however, do not greatly concern us, although it is of interest to know that such steels are practically non-expansible. Principal uses of ordinary nickel steels are for drop forgings, case-hardening parts, and general structural purposes.

When nickel steel is cooled gradually, its structure is pearlitic. In this condition it is capable of various heat-treatments designed to produce specific series of mechanical characteristics. It is more readily forged than carbon steel proper, and has greater strength without any corresponding loss of ductility.

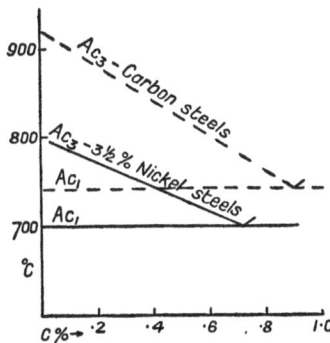

Fig. 16.—Showing effect of 3·5 per cent of nickel on the critical points of steel

An important feature of nickel is that it does not permeate the steel very quickly, as chromium does, and therefore the crystals or grains do not increase in size with anything like the same rapidity. Nickel, in fact, produces steels of fine grain, and fine grain is often synonymous with toughness.

Overheating of a nickel steel is not fraught with consequences quite so disastrous as is the case with chromium steels. This does not, of course, mean that these steels should be heat-treated carelessly.

Tungsten Steels.

Tungsten steels are principally employed in the manufacture of steels for tools. This element has the advantage of refining the crystalline structure of steel, producing small-sized grains. Tungsten in small quantities forms a carbide which is very stable. When the high-speed cutting steels containing from 14 to 22 per cent of the element are examined, however, another constituent appears, namely, an extremely hard iron tungstide. The commercial tungsten steels contain from 0·5 to 22 per cent of the metal. Characteristic uses are for dies, springs, taps, permanent magnets, drawing dies, and all forms of machine cutting tools.

Tungsten raises the temperatures at which structural transformations on heating occur. The cooling changes, however, are greatly retarded, so that the tungsten steels are strongly air-hardening. It takes a very much higher temperature to decompose the extremely hard martensitic structure than is the case with carbon tool steels. It is this which renders tungsten high-speed steel so satisfactory a cutting steel. When a tool is taking a heavy cut in the machine, great heat is generated at the tool nose. The heat is high enough to break down the martensitic structure of a carbon steel tool, causing immediate failure to cut, but not that of a tungsten high-speed steel tool, which will go on cutting even at a red heat.

Tungsten steels are still used to a minor extent for permanent magnets, but have been largely replaced by cobalt, nickel-iron-aluminium, and nickel-cobalt-aluminium-iron steels. They only give their best magnetic properties when fully hardened, and the magnetic values are much lower than those given by the more recently discovered alloys.

The structure of high-speed steel is distinctly intricate. Double carbides appear—tungsten carbide, chromium carbide, &c., and also iron tungstide. These carbides, however, will dissolve each in the other even in the solid condition. Nevertheless, although the solid solution thereby created is a carbide of very complicated type, its appearance under the microscope is structurally simple. It is easy, therefore, to treat this carbide as if it were one substance. Another point to be noted is that the iron tungstide will form a solid solution with iron. These facts combine to simplify the metallography of the high-speed steels.

We have briefly run through the structural and physical characteristics of the simpler and common alloy steels. As already indicated, it would be possible to pursue the subject indefinitely, ranging through the molybdenum, vanadium, nickel-chromium, and other complex alloy steels. This, however, would demand a more comprehensive treatise, and would take us too far from the essential subject, which is the *structure* of steel.

In the next chapter we will revert, therefore, to structure, and will examine in greater detail certain aspects of the subject which have hitherto been touched on but lightly, namely, crystal formation and growth, and the numerous constituents of which we have made mention, e.g. ferrite, martensite, &c.

CHAPTER X

Microscopic Structure

Behaviour of Steel on Cooling.

In this chapter we are going to study more closely the behaviour of steel on cooling as regards changes in its structure. This means that we must revert, in imagination, to our etched and polished specimen under the microscope. What happens when steel begins to solidify? The late Dr. Rosenhain studied this subject and considered that the crystallization of a metal could be represented by means of a set of children's building bricks, built up to cover a given area but starting from different points and building in different directions, as represented in figs. 17 to 21. These show in a clear and interesting manner the birth of a metal crystalline structure such as iron, and the method by which it can be recognized when microscopically examined.

Fig. 17 shows what occurs on the first beginnings of solidification. The main bulk of the metal is still fluid, but a few grains or crystals have become solid and have here and there joined up in tiny aggregations.

In fig. 18, cooling has proceeded, with the result that more crystals have formed and attached themselves to the previous bands.

Figs. 19 and 20 represent further stages in the cooling, the crystals growing larger until, as represented in fig. 21, the entire mass has become completely solid. In figs. 17 to 20 the structure of the solid matter is represented as minute aggregates of true cubic form. When the final stage of solidification is reached, however, each set of cubical blocks interferes with the others, so that if the last bricks are to be made to fit the intervening spaces, they must become distorted. These distorted bricks then constitute the lines shown in fig. 21. What it really means is that the atoms of the metal which are

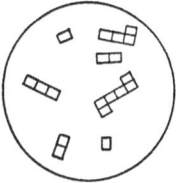  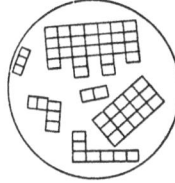

Fig. 17 Fig. 18 Fig. 19

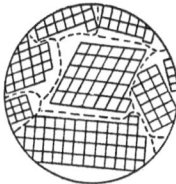 

Fig. 20 Fig. 21

last to freeze cannot possibly be arranged on a regular space-lattice, but are distributed at random, as it were. The lines in fig. 21 really represent, therefore, non-crystalline material formed between adjacent crystal grains.

Preparation of Specimens for the Microscope.

We have referred in an earlier chapter to the fact that specimens of steel have to be prepared specially for examination under the microscope. The attentive reader may have wondered why this is. Surely any piece of steel under the instrument would show its structure quite plainly? The answer is that it doesn't!

First of all, a microscope specimen must be perfectly flat. If there is the slightest convexity or concavity, it will be difficult, if not impossible, to focus the entire field properly, and either the image will be distorted, or the structure will not be visible in its entirety. Secondly, a merely flat (in the non-technical sense) specimen will not in itself suffice. It must be so flat that no ridges or burrs or inequalities appear, to distort the image. Thus, to saw a piece of steel from a bar with a hacksaw will not do. The surface will not be even enough. The metallurgist has, therefore, to file it first, or for preference, grind it mechanically, using a very smooth emery wheel. But even then, the file or wheel will leave behind minute excoriations or scratches, quite deep enough to show up under the microscope and prevent the formation of a clear and accurate image. The next step, therefore, is to take out these scratches by rubbing the specimen with emery papers, beginning with a coarse, and passing by stages to the finest, grade. The emery paper is placed on a smooth piece of glass of fair thickness.

But even when the finest emery paper has done its work, the specimen is still not suitable for microscopical examination. To remove every, even the minutest, abrasion, it has to be polished on a rotating block covered with a suitable cloth. This cloth is damped with water and coated with some very fine polishing powder, e.g. jewellers' rouge or powdered alumina. Emerging from this last operation, the specimen has a perfect mirror-polish. The untrained student, relieved to see the end of his labours, washes and dries the specimen and places it under the microscope, applies his eye to the eyepiece, and prepares to see the marvels of steel structure.

Etching and Mounting.

Alas! He sees no structure at all (unless he has somehow got hold of one of the exceptions to the rule). He consults his tutor, and learns that something more has to be done before the steel's structure becomes visible. Why is this? The prevailing theory is that the polishing operation creates a kind of surface layer or film which is not of crystalline character. The crystals on the surface have, so to speak, flowed together to create a uniformly spread out external film. Before the

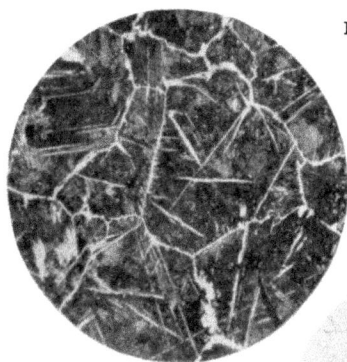

Fig. 22

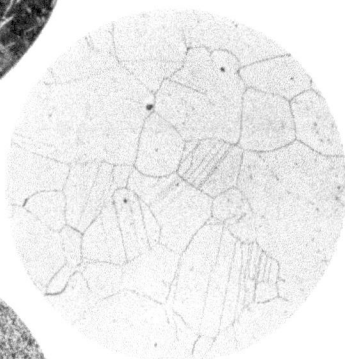

Fig. 23

Fig. 24

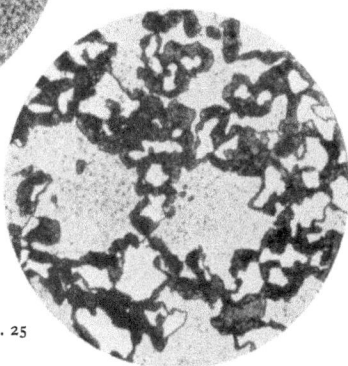

Fig. 25

actual crystalline structure corresponding to the condition of the metal can be seen, this film has to be removed. The quickest and easiest way of doing this is by etching, which means eating it away with the aid of a suitable acid. Usually 2 per cent nitric acid in alcohol or a saturated solution of picric acid in alcohol is used.

It must be mentioned here that with pure metals and solid solutions, a number of crystals are usually so positioned that in respect of their fellows they become electro-positive, and are differentially attacked by the etching fluid or reagent. The effect of this is to cause a slight difference in colour as compared with the other crystals, though they may be of identical type. Different acids are used to bring out different structures.

Washing is essential after etching, to remove every drop of the acid and prevent complete surface discoloration. The specimen is then dried with alcohol. Sometimes it is also washed with alcohol in order to produce clear marking of the constituents.

The next stage is to mount the specimen in the most suitable way for examination under the microscope, though this is not always essential, since some of the elaborate modern photo-microscopes will take specimens of quite large dimensions. Usually, however, the specimen is attached to a small plate or strip of ground glass by means of an adhesive substance not too rigid in character, e.g. plasticine.

Microscopic Study of Structure.

We are now in a position to study the structure microscopically. (It is assumed that the reader needs no special instruction in the use of the microscope.) The first thing observed will be a difference in colour between the parts which have been attacked by the etching acid and those which have not. The former will be dark and non-reflecting, the latter bright and reflecting the light they receive.

If our specimen is of pure iron, it will be found that it is composed of a number of crystals of many-sided character, as seen in fig. 4 (p. 14). This is the typical structure of *ferrite*. Quite a large number of metallic elements, such as manganese, nickel, silicon, chromium, can be dissolved in solid alpha iron, forming solid solutions. These solutions will show under the microscope structures very much like the ferrite. The technical term "*ferrite*" is used, therefore, for any solid solution of an alloying element in alpha iron.

In examining the ferritic structure we observe certain fine lines which separate the crystals from each other. These are the crystal or grain boundaries. These lines are visible because the etching reagent has attacked the crystal edges just enough to produce a slight darkening of tint.

If, in place of pure iron, we substitute a specimen of mild steel (less than 0·89 per cent C) a different form of structure is seen, as in fig. 5 (p. 14). Here certain portions are light and are obviously akin to the ferritic structure just examined. They are, in fact, ferrite. But dotted here and there among these lighter crystals are patches of darkly-tinted matter. This darkly-tinted stuff has to be magnified still more highly before we can see what it is. It is then discovered that it is not one single ingredient, but a combination of two (a eutectoid, in fact). It is a combination of alternate streaks, layers, films, or, to use the metallurgist's term, lamellæ, of iron (ferrite) and carbide of iron (Fe_3C). This material is technically known as pearlite, because under a low-powered microscope it often presents the appearance of mother-of-pearl.

If we take a series of specimens in which the carbon content is progressively raised, it will be found that the amount of pearlite steadily increases, the ferrite being correspondingly reduced, until we reach the eutectoid of 0·89 per cent carbon, when the structure is completely pearlitic, so that pearlite can be regarded as a substance containing 0·89 per cent carbon or 13·35 per cent iron carbide.

We shall next discuss cementite, austenite, martensite, &c., from the microscopical point of view.

Cementite: Pearlite: Austenite.

When a piece of steel contains more than 0·90 per cent of carbon, i.e. when the eutectoid point is passed, a new constituent is revealed in its structure by the microscope. This constituent is free carbide of iron, or *cementite*. Unless special etching reagents are used, such as a boiling solution of alkaline sodium picrate, it is not always possible to distinguish cementite from *ferrite*. But if the reagent referred to is used, the cementite will appear coloured, and the ferrite will remain whitish. If a commoner reagent, such as nitric acid, is used, the cementite will usually be whiter in appearance than the ferrite. Fig. 22 shows cementite and pearlite in a photomicrograph of steel.

It has already been explained that *austenite* is a solid solution of carbide of iron in gamma iron. It has also been shown that when a piece of steel is quenched and then tempered it may go through a series of successive changes from austenite, its original form, to *pearlite*. Austenite, whose structure is shown in fig. 23, has been fully dealt with earlier.

Martensite: Troostite: Sorbite.

Martensite is the first stage in the breakdown of austenite and its structure is shown in fig. 26. The needle-like intersecting shapes are quite easily recognized and are characteristic of martensite. It must be borne in mind, however, that martensite does not always appear as a needle-like constituent. Sometimes it appears as an apparently structureless ivory-tinted constituent when the specimens are lightly etched with picric acid.

From martensite we pass on to *troostite*. This constituent is very readily attacked by etching acids, and turns dark brown or black. Fig. 25 shows troostite surrounding areas of ferrite, with martensite in between. Troostite contains particles of iron carbide which, however, are too small for the microscope to reveal. It is formed when the cooling speed is not quick enough to produce an entirely martensitic structure, as when small pieces of carbon steel are oil-quenched, or in the interiors of larger pieces water-quenched. It is really an exceedingly intimate mixture of ferrite and carbide.

The next constituent is *sorbite*, often said to be formed when the cooling speed lies between that needed to produce troostite and that needed to produce pearlite, but more commonly formed by the full tempering of quenched or hardened steel. Its structure is shown in fig. 24, from which it will be observed that it is an intimate mixture of ferrite and carbide in which, however, unlike troostite, the carbide particles are readily observed.

Influence of Annealing.

One other important feature of the microstructure of steel is shown in figs. 27 and 28, which serve to illustrate the influence of annealing on the structures of mild and medium carbon steel castings. Annealing destroys the cast (or so-called Widmannstätten) structure, replacing it by a more rounded and uniform structure; in consequence the ductility and toughness are enormously improved.

Macrography.

The microscope is not the only weapon the steel manufacturer possesses in his struggle to know more and ever more about steel and its structure. Valuable as the microscope is, it has one serious limitation. It reveals only one small piece or section of the steel under examination, and it has to be assumed that the whole of the piece is uniform and corresponds in structure to that of the specimen etched for microscopical investigation. This assumption is not always justified, and if the evidence of structure provided by the microscope conflicts with the facts as given by performance or mechanical test, some further steps must be taken to ascertain if the steel really is uniform in structure with the microsection taken from it.

Fig. 26

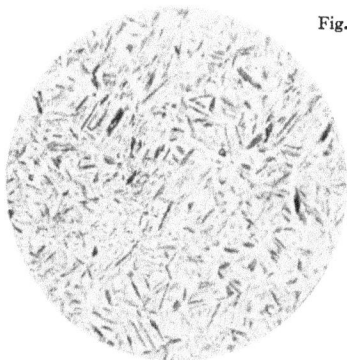

Fig. 27

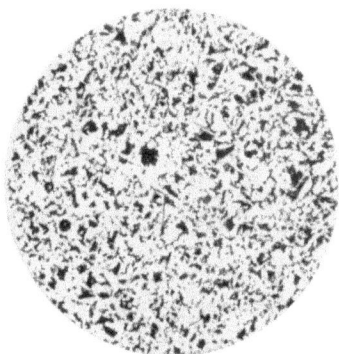

Fig. 28

Fig. 29
(Microphotograph,
8 magnifications)

Facing p. 72

A whole branch of metallurgical inquiry known as macrography at once presents itself, and must be alluded to in these pages. As compared with microscopy, macrography deals with the whole mass of the steel, whose structure from this point of view is then known as its " macrostructure ". Microscopy, on the other hand, concentrates on an extremely small section of the metal. Macrography will not, therefore, be so precise a means of investigation as microscopy, but it will enable us to decide with some accuracy how far the macrostructure of the steel is homogeneous.

The samples for macroscopic examination have to be polished and etched just as with the microscope, but this preliminary work will be less exacting, because it is not or seldom possible to deal with a large mass as thoroughly and meticulously as with a small microsection. The metal surface is first planed, then ground mechanically or hand-filed. Polishing with coarse and fine emery cloths follows. No local heating, either during planing, grinding, or polishing, should be allowed. The process completed, the prepared surface is rubbed with cotton-wool steeped in alcohol, for the purpose of eliminating oil, grease, &c. Etching then follows.

The etching reagent usually employed is 10 per cent nitric acid or a copper reagent, and the period of etching varies with the analysis of the steel. After being etched, the surface is washed thoroughly with hot water, rubbed with cotton-wool, then cleaned and dried with benzol. A final light rubbing with No. o emery paper is recommended by Gregory for the purpose of bringing out the structure more plainly.

The surface is then photographed, or, alternatively, it is coated with printer's ink and, by means of a dry rubbered roller, the impression is transferred to a sheet of art (shiny-surfaced) paper.

Sulphur Print.

Another method is that of the sulphur print. In this, a sheet of bromide paper is steeped thoroughly in 5 per cent dilute sulphuric acid, and laid on to the polished prepared surface. The rubbered roller is employed to squeeze out any air-bubbles. The manganese and iron sulphides decompose under the action of the acid, and give off sulphuretted hydrogen. This reacts on the silver salt in the bromide paper, and produces a dark brown stain of silver sulphide. After having been left in position long enough for these chemical reactions to occur, the paper is peeled off, washed with water, and fixed in the ordinary solution of hypo. Exactly what is then revealed is an impression of the regions in which the sulphides lay.

What the steel maker learns from macro-etching and from sulphur prints will be outlined in the next chapter.

CHAPTER XI

Testing of Steel

Macrography: Testing.

The macrophotograph is primarily designed to reveal the structure of an ingot or other large piece of steel, and it is specially important as a means of indicating the " flow " of the metal in parts that have been forged or rolled, used, strained, or worked in other ways. In addition, this and the sulphur-print method are valuable as revealing the manner in which various impurities are distributed. Local impurities or segregations are clearly shown, and examination of the macrographic structure is a handy check on the routine or special analysis of the steel itself. Analysis may be satisfactory because the specimen has been taken from an area in which no segregated impurities occur, whereas the macrophotograph or sulphur print may show marked segregation at a particular point. Macrography is also useful in connexion with welded or soldered joints, revealing the presence or absence of impurities.

To read a macrophotograph with accuracy is not possible without considerable practice. Fig. 29 shows a typical macrophotograph, and fig. 30 a characteristic sulphur print.

To determine the structure of steel by microscopy and macrography does not, however, completely meet the requirements of the steel manufacturer and metallurgist. The microscope, as we have seen, focusses attention on minute constituents, and by its aid a wonderfully interesting science of microstructure and its interpretation has sprung up, and has been invaluable. Similarly, macrography concentrates attention on the general structure of a steel, and reveals hitherto unknown characteristics of its structure and composition.

Both these methods are based, nevertheless, on the appearance of steel. There is a third method by which the inquirer is enabled to determine with some accuracy the structure and characteristics of a given steel. This method is based on the way steel *behaves* as compared with how it looks, and it covers all that portion of metallurgical investigation known as mechanical testing.

It is important to remember that none of these three methods is adequate in itself. Each tells the skilled inquirer a great deal, but it does not tell him all. The three together, or a combination of two, are necessary if a comprehensive and intelligible picture of the steel's structure is desired. One of the greatest factors in the advance of modern steels has been, in fact, the development and perfecting of microscopical and other methods of examination.

The Tensile Test.

The commonest mechanical test is the *tensile test*. It is simple and readily understandable. If you take a steel rod and pull it, holding an end in each hand and exerting your pulls in the direction of the rod itself, as you would pull a piece of cotton in order to break it, you will—if you possess the strength of a Hercules—gradually cause the steel to stretch until, ultimately, it breaks. The tougher the steel, the more force will be needed to break it. In substance, that is the tensile test, except that for the bare hands is substituted a machine, whose pulling power is registered exactly, and for the rod a carefully machined test-piece, whose dimensions are measured both before and after the test. The pull exerted by the machine on the specimen is known as the " stress ", and when the test-piece finally fractures, the maximum load that has been applied to bring this about is carefully recorded. But as we also know the cross-sectional area of the test-piece, it is possible to calculate how many pounds or tons per square

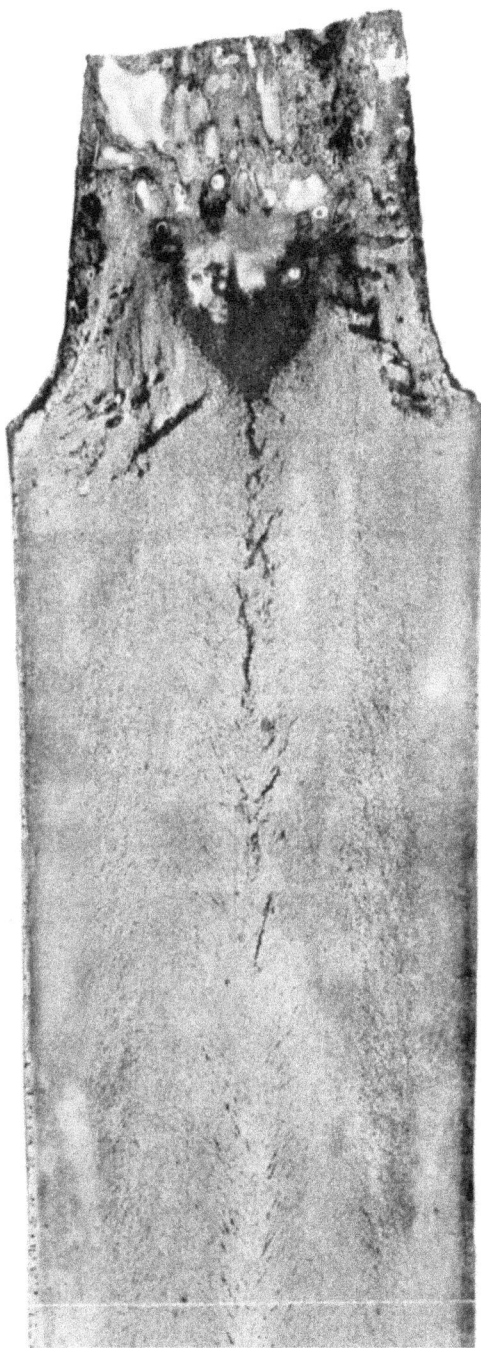

Fig. 30.—Photograph of sulphur print of 2-ton ingot. V-segregates clearly shown. Approx. ⅓ actual size

inch of load or stress can be withstood by the steel before it breaks. The result is called the tensile strength of the steel, or, more accurately, the ultimate strength or *maximum stress* of the steel in tons per square inch. When, therefore, a steel is described briefly as 40-ton steel, what is meant is a steel capable of withstanding an applied load or stress of 40 tons per square inch before it breaks.

Tensile Strength: Elastic Limit: Yield Point.

But a pulling or tensile test such as has been described tells us rather more about the steel than this. It tells us what is called the *elastic limit* of the steel. Now one of the peculiarities of steel not hitherto described is that it is to a certain degree elastic. In other words, if you pull it as you would pull a piece of elastic, and do not pull too hard, it will stretch and as soon as the pull is removed, spring back to its original length. But if you go on by degrees, pulling harder each time, and then letting go, you eventually come to a point at which the steel does *not* spring back, but remains stretched. It is definitely and irrevocably longer. It has, in metallurgical language, passed its elastic limit.

The machine in which the tensile test is carried out registers the exact load required to produce a permanent set in the steel, in tons per square inch. We have thus ascertained from the one test both the tensile strength and the elastic limit of the steel. But there is a third property which can also be ascertained. This is the property of elongation before fracture that the steel possesses. This needs some explanation.

When a steel test-piece is loaded beyond its elastic limit it continues to stretch, but if now the load is taken off, it will no longer return to its original dimensions, i.e. it will have acquired a certain amount of permanent elongation or permanent set. If, however, we do not remove but go on increasing the load, we find that the lengthening of the test-piece proceeds at a much faster pace in relation to the increase of load than was previously the case. To express this phenomenon scientifically, we say that the strain increases more swiftly than the stress. This goes on until a point is reached at which there is sudden very great elongation (or increase of strain) produced by a very small addition to the load. This point of rapid strain increase is known as the *yield point*.

This growth of strain in relation to stress is often illustrated by means of a diagram, known as the stress-strain curve (see

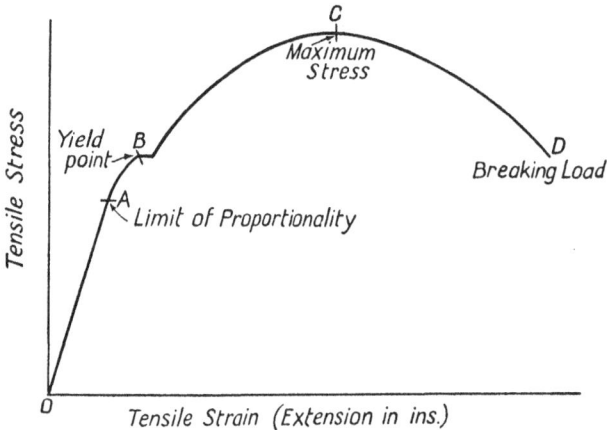

Fig. 31

fig. 31). From this it will be seen that from the moment when the elastic limit or point of permanent set is reached, the strain increases proportionately to the stress until the yield point is reached. After the sudden yielding at this point further extension occurs and the maximum stress (C) is reached, after which the piece stretches until finally it breaks.

Elongation: Reduction of Area.

Now having broken the test-piece in the manner described, the steel manufacturer wishes to measure this property of *elongation*. The standard two inches of test-piece are taken, and measured after fracture. From this measurement it is quite easy to work out the percentage of elongation. If the

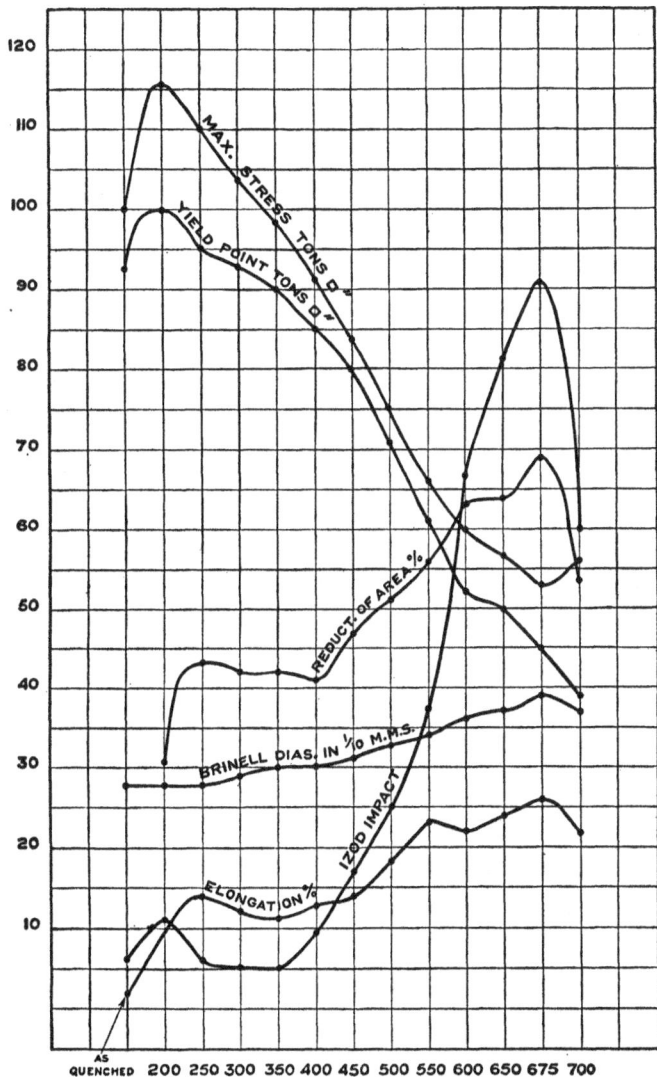

Fig. 32

test-piece before testing measures 2·0 in. and after fracture measures 2·2 in., it is obvious that the elongation has amounted to 10 per cent.

But the test tells us something more. To pull an elastic rod until it breaks not only lengthens it, but also makes it thinner, or, in other words, reduces the area of a section across it. By measuring the *smallest* cross-section after breakage and comparing this with the original area, it is easy to calculate the *reduction of area* expressed as a percentage. Both elongation percentage and reduction of area percentage are useful as a guide to the steel's ductility.

Fig. 32 shows the effect of heat treatment on the mechanical properties of a nickel chrome steel for gears, quenched in oil at 850° C., and tempered for 40 minutes at the temperatures indicated (degrees Centigrade). The figure brings out clearly the effect of faulty heat-treatment in the earlier parts of the elongation and impact curves (see fig. 32), and other effects can also be deduced.

Fig. 33 shows for comparison with each other, the load-extension curves for a mild steel and for a nickel chrome steel.

We now go on to compression, bend, and impact tests.

Compression Test.

A form of test of importance to constructional engineers is the *compression test*, whereby the ability of steel to resist the immense compressive force developed by a heavy weight resting upon it is ascertained. A New York skyscraper will serve as a typical example. The steel framework of this is pressed upon by the enormous weight of the structure (such as floors and roofs). This compressive force is transferred to the rock on which the building is founded. The resistance of the rock exerts a counter force, and if the steel were unable to withstand the combined action of these two forces, it would be squeezed until it collapsed.

In designing a skyscraper, an engineer must, therefore, have some means of knowing what compressive force the steel he

is going to use for his main framework will stand. For this
reason tests of compression have been devised, and the tensile
testing machine can be used for carrying them out, after a
number of modifications have been made. The difference is,

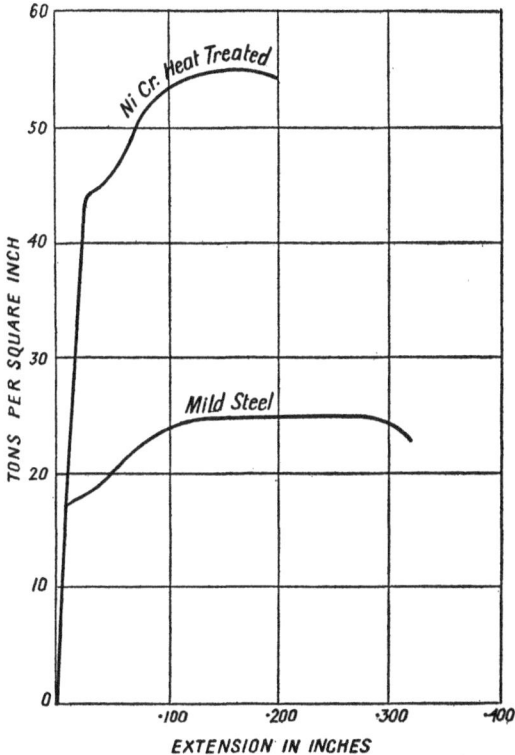

Fig. 33

of course, that instead of tugging mechanically at the test-
piece, you exert pressure of a compressive character until the
steel " gives " and will not recover. In other words, there is
an elastic limit and yield point for compressed steel as there
is for elongated or pulled steel. These values may be used by

the engineer, but it must be realized that the figures obtained relate to compression and not to tension. The result is, likewise, expressed in tons per square inch. One peculiar feature demands attention. It might be supposed that the compression-resistance would be equal to the resistance to tension. This is not so. The steel will actually stand more compression than it will tension, the proportion being between $1 : 1\frac{1}{2}$ and $1 : 2$.

Bend Tests.

Bend tests are another important type of test for steel. They can be divided into two kinds, the simple and the complex. The simple bend test is merely a method of determining whether a steel is adequately ductile, and consists in actually bending the piece by mechanical means to the required degree, without any special measurement of the force exerted. In the bend test proper, however, the test-piece is of specified dimensions, supported at the ends, and subjected to a bending force applied to the centre, or alternatively bent about a fixed radius. If fracture occurs before the given degree of bending has occurred, the material has failed to pass the test. It may be enough for the user to know that the steel will bend through 90° or 180° without fracture. On the other hand, he may desire to know exactly what amount of force is required to bend the steel until it breaks, in which case the test is continued until fracture occurs, when the necessary measurements are made as in the tensile test.

Impact Test.

The next test of importance is the *impact test*, the most popular form of which in this country is the Izod. It is sometimes known as the notched bar test. An alternative method is the Charpy test.

In the *Izod* impact test, the machine consists of a heavy base with two standards bolted on, which support the pivot of a pendulum. The pendulum itself swings in ball bearings and develops an energy of 120 foot-pounds. The test-piece, having a notch of standard size, is held in a vice at the base, with the notch facing the pendulum, and the pendulum strikes it with a hardened steel knife edge. In breaking under the shock of the impact, the test-piece absorbs a certain amount of the energy of the pendulum. Measurement of this is obtained by the continued and diminished swing of the pendulum, which moves an idle pointer across a graduated scale. The greater the resistance of the steel, the less the continued swing of the pendulum, and, therefore, the shorter the distance covered by the pointer. The impact figure is obtained in foot-pounds by comparison of the distance travelled by the pointer with a table of values carefully compiled.

This test is used in order to determine the ability of a steel to resist shock. This makes it extremely valuable in the testing of steels for automobiles and aircraft, where inability to withstand a sudden shock might have disastrous results.

Hardness: Brinell and other Tests.

One now comes to tests of *hardness* or resistance to penetrative action caused by pressure. It is necessary to point out in this connexion that no absolute test of *hardness* exists, but a number of relative tests are employed, the most usual being the Brinell ball test.

The method adopted is to force a small hardened and tempered chrome steel or tungsten carbide ball 10 mm. diameter under pressure into the steel to be tested. The apparatus is operated by oil pressure. The steel for test is placed on a platform which is raised by a screw until in contact with the ball. Pressure is then applied by means of the oil pump. The pressure on the specimen is read from the manometer attached to the machine. A controlling device in the form of a balance is also fitted. The centre pin of this rests on the balls in the cylinders, which serve as a piston, and carries weights, so that the pressure on the ball may be varied according to the hardness of the material to be tested. When the standard pressure of 3000 kg. is being exerted, these weights float.

The valve is left closed for 15 sec. and then released. When the pressure relaxes, the test specimen may be withdrawn. It is then put under a microscope, the inner lens of which carries a millimetre scale divided into tenths. This enables the diameter of the ball impression to be accurately measured.

The Brinell test is, of course, of no use for thin sheet metal, nor for thinly case-hardened steels.

The hardness number is ascertained by measuring the diameter of the impression or dent made by the ball, computing the area of this dent, and dividing the standard load by this figure. Obviously, the harder the steel, the smaller the dent and, therefore, the higher the ultimate Brinell number.

It has been established that there is a definite relation between Brinell hardness and tensile strength, enabling the one to be calculated from the other. Table I, p. 110, illustrates this relation clearly. Sometimes a load of less than 3000 kg. is used with a ball having a diameter less than 10 mm. The hardness values can then only be compared when the ratio of the load divided by the square of the diameter of the ball is maintained constant. Thus the standard ratio is 30 : 1 (10 mm. ball and 3000 kg.), so that a ball of 1 mm. diameter with a load of 30 kg. gives standard Brinell hardness values.

There are numerous other methods of giving a value to hardness, among which may be mentioned the Rockwell, Scleroscope, Firth, Vickers diamond, Herbert Pendulum, Monotron, Duroskope, Durometer, Drop Hardness, and Moh. In Table II, p. 111, a conversion table of various hardness numbers is given, and this, if referred to, will serve as an indication of the relation of the various figures to one another.

Before finally leaving the subject of tests, we shall briefly examine now such tests as have not yet been dealt with.

Torsion: Shearing: Fatigue.

Tests for steel are numerous, and the details thus far given by no means exhaust the number. There are, for instance, *torsion tests*, which indicate the ability of the steel to withstand twisting stresses; *shearing tests*, which ascertain the ability of steel to resist a shearing action (as with rivets, which are specially subject to this form of stress); and repeated stress tests, which are really *fatigue tests*. In this last type, the test-piece is made to suffer first a pulling stress, then a pushing stress, each repeated alternately in quick succession and for a long spell. It is well known that quickly repeated stresses are more likely to set up fatigue and eventual fracture than a continuous stress applied in one direction or sense only.

An interesting example of quickly repeated stresses causing fatigue fracture is the crankshaft of an engine, in which at each stroke of the engine the twisting and bending of the shaft alternate with the loads on the piston of the engine.

Another example is the axle of a railway wagon, in which the stresses change from compression to tension with each revolution of the wheel.

Dilatometric Test.

Another test, which is of real importance, is the *dilatometric test*; it indicates the volume changes which take place when steel is heated or cooled.

As we have seen in previous sections, heating or cooling steel leads to modifications in the volume of the steel and also in its structure. These the test in question measures and records. It is necessary to know in advance the particular factor it is desired to measure. The coefficient of cubical expansion being known, it is possible to discover the variations per unit length and volume, from which in turn the specific gravity and density can be ascertained. These various factors are all related to one another, so that it is possible to measure expansion when temperature is the variable or the density, when the analysis of the steel is a variable.

(F 385)   7

It is the dilatometer that has enabled metallurgists to acquire a good deal of their knowledge of the exact change-points of steel as regards both time and temperature, and also of the actual meaning of these changes. In the ordinary way it is difficult, if not impossible, to locate with any exactitude the transformation points of steels. Even if a fair approximation

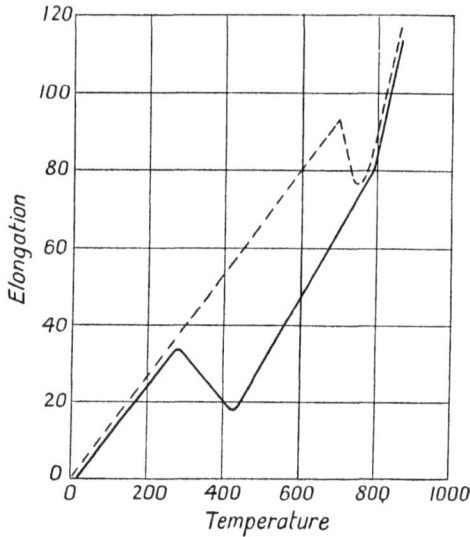

Fig. 34

is attainable at all, it can only be got by reducing the cooling speed of the heated steel to a minimum.

The dilatometer, however, enables the metallurgist to measure with exactness changes occurring within very close limits of temperature, so that the phenomena accompanying the transformation of structure can be recorded immediately. The record takes the form of a curve, such as that shown in fig. 34, due to Grenet, for ordinary stainless steel. Examining this curve, one should note that the dotted line represents the heating curve and the unbroken line the cooling curve. The reader will note that the steel, stable at room temperatures, increases in volume with complete regularity up to 660°. The change-point then appears, the steel then actually contracts owing to the alpha-gamma iron change, but at the temperature of 685° the expansion begins again and continues as the temperature is raised further.

Taking the cooling curve, it will be seen that reduction in volume due to contraction is at first quite regular, until the change-point at 410° occurs.

The advantage of the dilatometer is that it enables the steel manufacturer to determine with exactitude, from an examination of the dilatation curves corresponding to the particular steels in which he is interested, the precise quenching and tempering temperatures for these steels. He knows the change-points exactly, and can so treat his steels as to produce the characteristics he requires within their given range.

Mechanical Characteristics of Steel.

Throughout this book emphasis has been laid on structure, and every attempt has been made to indicate the types of structure and their theoretical significance in regard to steel. But the significance of the structure is practical as well as theoretical. Every impurity, every variation of carbon or alloy content, every differing structural condition brought about by differing forms of heat or mechanical treatment, has some effect on the properties and performance of steel in use. It is necessary, therefore, to outline as briefly and simply as possible what these variations signify in relation to the mechanical characteristics of steel.

Carbon Percentage.

In ordinary steel, the mechanical properties are chiefly affected by modification of the carbon percentage. As soon as carbon is introduced into iron the tensile strength begins to rise from about 20 tons per square inch up to approximately 60 tons per square inch with normalized steels, which figure is attained with a carbon percentage of about 1·0. If carbon continues to be added, the tensile strength does not, as one might imagine, go on increasing, but actually declines. What does this mean?

It has already been shown that 0·9 per cent of carbon is the eutectoid or pearlitic condition. 0·9 per cent is near enough to 1·0 per cent to make it evident that there is a possible

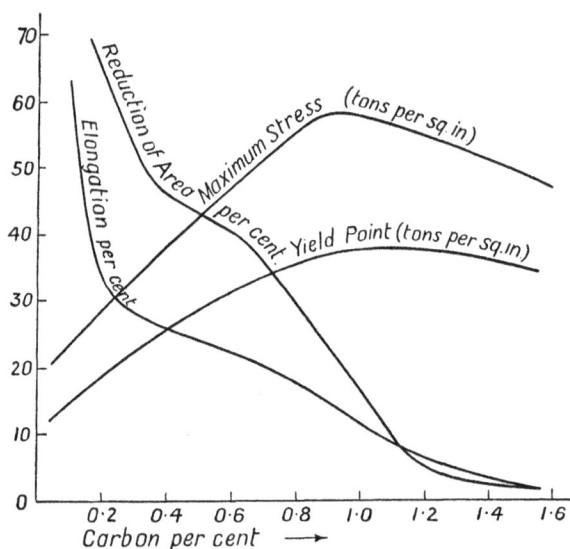

Fig. 35.—Showing relation between mechanical properties of normalized steels and their carbon content

relation between the eutectoid point and the tensile strength of the steel. Now tensile strength, being the resistance of steel to a pulling stress, cannot be consistent with ductility, which is the ability of a metal to be hammered or pressed out into shape without cracking. Any increase in tensile strength due to the addition of carbon will be accompanied by a decrease of ductility. Virtually, this means that with increased tensile strength the hardness increases. Fig. 35 shows, in graphic form, the relation between the mechanical characteristics of normalized steel and the carbon content.

Impurities.

Turning now to the two principal impurities of steel, one may deal first with sulphur, which is seldom allowed to become high enough in percentage (above 0·10 per cent) in steel to affect its mechanical properties to any marked degree. If it were, the steel would be embrittled. Phosphorus is the most harmful impurity in steel. It causes brittleness and for this reason is generally kept below 0·10 per cent in acid steel and 0·05 per cent in basic steel. It has been ascertained that up to 0·10 per cent of phosphorus may actually increase the tensile strength of steel to a slight extent, after which the strength declines again, but this is no argument for increasing the phosphorus content, because the steel is unable to resist sudden shocks adequately if the phosphorus content is increased beyond the limits stated.

Heat-Treatment.

Heat-treatment is the third factor affecting the mechanical characteristics of a steel. There are various forms of heat-treatment, and each of these modifies the mechanical properties to some extent. A consideration of these effects is therefore essential at this point.

Heat-treatment is divisible into four sections: (a) hardening, (b) tempering, (c) annealing, and (d) grain refining (normalizing). Taking hardening first, with a eutectoid steel (0·9 per cent carbon) as the basis, it will be found that hardening or quenching from a temperature above the critical point produces a martensitic structure which corresponds to a steel so hard as to be capable of scratching glass. Such a steel is, however, extremely brittle and possesses little ductility. As the carbon percentage is lowered, the steel's hardness, after quenching from above the upper critical point, decreases. Again, the higher the temperature from which the steel is

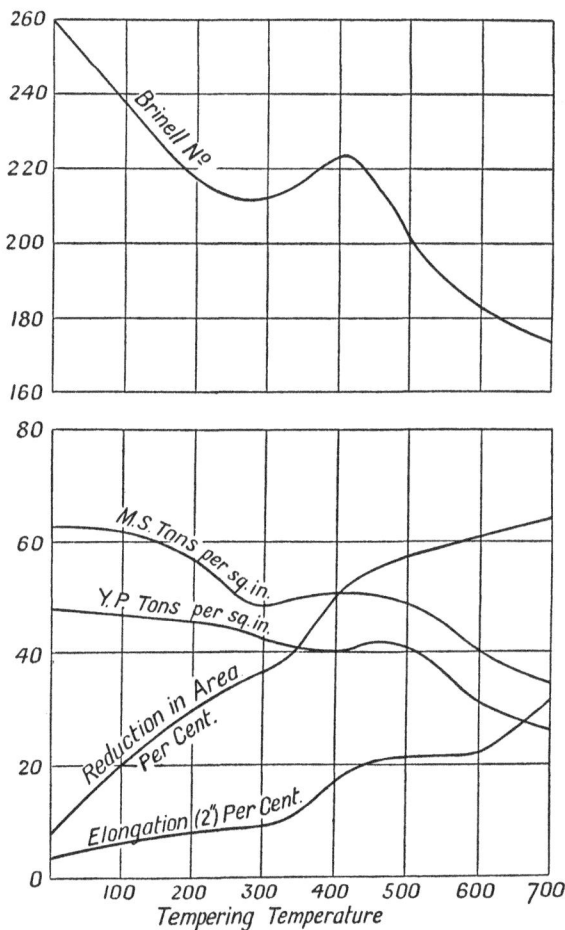

Fig. 36

quenched, the larger are the grains formed, and in consequence the greater the brittleness of the steel.

Tempering a hardened carbon steel at 200° C. somewhat
lowers the yield point and the maximum stress, raises the

elongation and reduction of area percentages, and reduces the hardness. By increasing the tempering temperature, a further reduction in tensile strength occurs, together with an increase in ductility, until at a tempering temperature of above 550° C. the tensile strength drops off perceptibly, if sufficient time is allowed, but the ductility is not simultaneously improved to any noteworthy extent. Fig. 36 shows the physical properties for various tempering temperatures of a 0·34 per cent carbon steel, based on figures quoted in Gregory's *Metallurgy*.

In practice, tempering temperatures of 600° to 650° C. are frequently employed in order to induce maximum toughness and ductility. The higher tempering temperature lowers the tensile strength, but ensures greater ductility.

Annealing is a softening process. Its effect, as shown in the following table for a 0·4 per cent carbon steel, is to lower the tensile strength and increase the ductility.

EFFECTS OF ANNEALING ON A 0·4 PER CENT CARBON STEEL
(From *Gregory's* " *Metallurgy* ")

Condition	Y.P. tons	M.S. tons	Elong. %	Redn. of area %	Brinell No.	Impact Ft.-lb.	Bend ¾" radius
As cast	23·7	43·2	8	7	229	3	25° broken
Annealed 900° C.	18·0	37·3	24	30	187	7	180° unbroken

Grain refining (normalizing) increases the tensile strength of an annealed steel (of equivalent carbon content); yield point is also raised; reduction of area and elongation are slightly reduced. Normalized 0·4 per cent C steel has a superior surface after machining to annealed steel.

CHAPTER XII

Corrosion: Stainless Steel

Rust.

One of the earliest things we learn about iron and steel is that they rust. " Rusty " iron is a very familiar material, and we must now consider why rusting takes place. It is actually a form of corrosion, and the rust itself is a hydrated oxide of iron, i.e. it is iron oxide which contains a certain amount of combined water.

It might at first be thought that rusting would most readily take place on those parts which are in direct contact with oxygen and water or water vapour. It can be shown, however, that oxidation (rusting) may take place more readily at points which are least accessible to oxygen. To account for this we must know something about the modern electro-chemical theory.

Electro-chemical Action.

On the basis of this theory corrosion or rusting is both an electrical and a chemical process. It is connected with the phenomenon known as *electrolysis*. When certain substances, e.g. acids and salts, are dissolved in water, the molecules of which they were initially composed are split up, partly at any rate, into negatively and positively charged particles known as *ions*. Because these ions are electrically charged the solution will conduct electricity, so that the solution is called an *electrolyte*. Now suppose we pass an electric current through the electrolyte solution. This marshals the ions into two groups as it were; the negatively charged ions are driven towards the positive pole (or anode) and the positively charged ions towards the negative pole (or cathode). The anode is the plate or point at which the electric current enters the solution, and the cathode is the plate or point at which it leaves the solution. When the ions reach the corresponding poles they are neutralized; they are no longer charged with electricity and those at the cathode are liberated or set free as elements. If the element is a gas, it comes away as such; if it is a metal it is deposited on the cathode. Thus if we start with a solution of very dilute sulphuric acid where we have hydrogen and sulphate ions in the solution, it is found that the hydrogen ions go to the cathode, where they are neutralized and converted into hydrogen gas molecules. Meanwhile the sulphate ions go to the anode, are neutralized, and then form oxygen gas molecules, sulphuric acid being re-formed at the same time. The net effect of the passage of the current is therefore to bring about the chemical decomposition of water.

Now suppose that we started, instead, with a solution of copper sulphate, and that both our anode and cathode were made of copper metal. In the solution this time there are copper ions and sulphate ions. When the copper ions reach the cathode they are neutralized, and copper is then deposited on the cathode. In the meantime the sulphate ions reach the anode, are neutralized, as before, but then dissolve copper from the anode to form copper sulphate. The net effect this time is therefore to transfer some of the copper from the anode to the cathode, i.e. the anode is dissolved and the cathode thickened up. This principle is applied in practical electro-plating where the articles to be plated, spoons, forks, &c., are made the cathodes.

Since, as we have just seen, the passage of a small electric current through an electrolyte solution is accompanied by chemical action, it is natural to assume that the reverse process will take place, and indeed it does, i.e. when chemical action takes place in an electrolyte solution, it is accompanied by the development of an electric current. Thus, if we place a rod of zinc and a rod of copper in a solution of copper sulphate, and join up the ends of the two rods to a voltmeter, the reading on the latter at once shows that a current of electricity is passing round the circuit. Actually what happens is this. The copper ions, as before, go to the copper rod, are neutralized there and deposited on the copper. The sulphate ions travel to the zinc rod, are neutralized, and then dissolve some of the zinc to form zinc sulphate. Copper is thus deposited and zinc dissolved or, in other words, the zinc becomes anodic to the copper rod, and the latter becomes cathodic to the zinc rod.

Rust: Iron and Steel.

Now we may have a state of affairs similar to this when iron and steel rust, where the electrolyte consists of water containing dissolved salts, or of condensed moisture which always contains carbon dioxide (carbonic acid gas) and, in industrial districts, the condensed water-vapour may contain sulphurous acid gases dissolved from the atmosphere. Most air at the seaside also contains particles of salt. In contact with impure water or condensed water-vapour some crystals of iron and steel become anodic to others, and are therefore corroded away. Impurities will behave in the same way, and we therefore have an explanation why " pitting " so frequently accompanies corrosion.

Pores or cracks in the surface may induce corrosion because the material at the bottom of these depressions or flaws becomes anodic, and the cracks or cavities are consequently enlarged. For these reasons *as smooth and well-polished a surface as possible is essential if a metal is to exhibit its greatest resistance to attack by corrosive influence.*

Further, it should now be clear why a material whose structure is uniform resists corrosion better than one having a non-uniform structure. The fact that pure iron resists corrosion and rusting much better than a mild or medium carbon annealed steel is well known, and a hardened carbon steel does not rust so readily as when the same steel has been annealed, because its structure is more uniform.

Film of Oxide.

We have mentioned that rust, which is a product of corrosion, is a hydrated oxide of iron. This means that as the iron dissolves it is converted into this form by the atmospheric oxygen dissolved in the water with which the iron or steel is in contact. At first the oxide coating is more or less uniformly distributed over the surface of the steel, but as it thickens up, it falls away and exposes a fresh surface for attack. Even if it does not fall off, however, the porosity of the oxide is such that it readily allows oxygen to pass through it, and the electro-chemical attack to continue. This is really the crux of the problem. Thus, if we could obtain a thin uniform coating of the oxide on the surface, which would not allow the substances in the corroding medium to pass through it, then a natural protection from further attack would be obtained, if the coating or film did not dissolve in the medium with which it was in contact. The remarkable natural resistance of aluminium to atmospheric corrosion is due to an exceedingly thin, but very tenacious, film of oxide on its surface. Unfortunately, where iron and ordinary steels are concerned, the iron oxide film first formed is neither tenacious nor impermeable, so that it does not afford any protection for the underlying metal.

Passive Film.

The ground is now cleared for a more detailed examination of corrosion and its phenomena. As has been seen, the conditions leading to the incidence of corrosion (damp or moisture, presence of corrosive substances, chemical composition, heat-treatment, deformation due to mechanical work) create an oxide film of very slight thickness which covers the metallic surface completely. Now research has proved that the extent to which corrosion proceeds or is arrested depends to a large extent on the character of the oxide film itself. This is an important point, as will be seen.

If the corrosion-producing conditions are such that the oxide film is not allowed to form, or if, as a result of the type of acid attack to which the steel is subjected, the film is dissolved immediately, there will be no protection afforded to the steel, which will immediately become the prey of swift and general corrosion.

Again, if the oxide film, by reason of its gradual permeation by metal ions, increases by degrees in thickness, the corrosive attack will be less rapid, but none the less general.

Thirdly, it may happen that although the film itself is intact, detached particles of corrosion products or of extraneous material alight upon it. These may cause a slight amount of local corrosion.

Sometimes the film will have suffered physical or local modifications which will give rise to the creation of oxygen concentration cells. This is enough to set up strong local corrosion.

Wherever the film has been eaten into by corrosive or erosive substances, whether fluid or solid, metal-ion concentration cells will be formed, and strong corrosion will ensue.

Finally, if certain chemical modifications render the film looser, so that it does not adhere as it should, there is more than a likelihood of a heightened corrosive attack, both local and general.

It is thus plain that the first essential as a preventive of corrosion is a protective oxide film which will not readily go into solution, will completely isolate the surface concerned from corrosive attack, and will not easily become loose or detached from the body of the metal. Such a film is termed a " passive " film.

Chromium and Resistance to Corrosion.

One remarkable fact is that the metal chromium is highly resistant to the corrosive attack of nitric acid. But before we discuss the chromium non-corrosive steels, let us consider this matter of passive films more closely. It is a singular phenomenon that if a piece of iron is steeped in concentrated nitric acid, or in certain acids containing chromium (e.g. chromic acid) it is not very greatly corroded, and if afterwards plunged into a different acid or other highly corrosive liquid that would normally have a strong corrosive effect upon it, it is found still to retain this relative immunity. The cause of this is that nitric acid is particularly favourable to the formation of a protective passive film, so that such degree of attack as it produces is, in a way, beneficial by virtue of the fact that corrosion by stronger reagents is minimized. Now if chromium, which is even more resistant to corrosion by nitric acid than iron, is alloyed with the iron, it is not to be wondered at that the resultant metal is very resistant indeed to nitric acid attack, a highly passive film being formed.

But the corrosion-resistance of iron-chromium alloys is not so simply explained as all that. When chromium, iron and carbon combine to make up an alloy of metallic type, a complex carbide is formed, in which the ferritic portion remains unchanged (so long as the steel is not hardened). In such alloys the chromium content is usually low. But steels of this type are not specially corrosion-resistant because the unmodified ferrite can readily be attacked by corrosive agencies.

It is only when the chromium percentage is raised to above 11·0 that structural modifications occur and render the steel resistant. This raising of the chromium percentage results in a surplus of chromium over and above what is required to form the double carbide of a low-chromium steel. The surplus chromium lies in solid solution in the iron and helps to form a chrome-bearing ferrite, the whole making an alloy offering high resistance to corrosion.

A most important point to remember is that if the carbon content of the alloy is increased, the chromium content must be correspondingly increased, or the resistance to acid attack will not be adequate.

Stainless Steel.

The happy idea of alloying iron and steel with chromium in an endeavour to produce alloys highly resistant to corrosive conditions was a natural consequence of such observations and conclusions, and has led to the production of the wide range of stainless steels which have done so much to brighten our civilization.

A stainless steel must have as a minimum 13–14 per cent of chromium to give it perfect freedom from corrosion, but nitric acid, ordinary attack by air and moisture, and organic acids of many types can be made harmless to a chromium-*iron* alloy if the carbon percentage is low enough to call for a lower percentage of chromium. If the carbon content is small enough, in fact, the entire chromium content goes into solid solution in the iron, forming a homogeneous alloy. This homogeneity will persist even when the metal has been normalized, and in many instances nothing more is necessary to make the alloy corrosion-resistant enough for its work.

A stainless steel, however, after subjection to annealing or normalizing, is not structurally homogeneous but heterogeneous, and to put it into the requisite homogeneous state it must be hardened. Hardening brings about uniformity of structure and simultaneously the greatest possible resistance to corrosion.

Stainless steel was discovered by Mr. Harry Brearley in 1913. The first stainless steels, which were mainly used for cutlery, contained from 13 to 14 per cent of chromium. Later it was found that, conversely, by reducing the carbon the chromium percentage could also be reduced, and a range of stainless irons produced, of malleable character, which would withstand the attack of many acids and ordinary atmospheric conditions.

The later development of the stainless steels and irons has been on the lines of the modification of their compositions by the introduction of nickel, silicon, molybdenum, titanium, tungsten, niobium (columbium), &c. In the main, however, there are two distinct classes of stainless steel: the straight 13–14 per cent chromium steels referred to above, and those containing large amounts of both nickel and chromium, such as the well-known " 18–8 " type. A straight chromium stainless steel of the cutlery type must be hardened by reheating to about 950° C. and quenching in oil, before it exhibits its greatest resistance to corrosion. In the hardened state its structure consists of martensite, its Brinell number is between 650 and 700, and the steel is magnetic.

The structures of the high-chromium high-nickel stainless steels, e.g. " 18–8 ", are essentially austenitic. They are, indeed, in the best condition to withstand attack when exposed to corrosive influence when their structures consist entirely of austenite crystals. This desired structure is obtained by heating the steel to about 1100° C., followed by quenching in oil or, with thin sections, by cooling in air. When thus treated the steel is non-magnetic and is soft, but, unlike the straight chromium steels, it offers its best resistance to corrosion when in this soft state.

Weld Decay.

Even with a good polished surface and correct chemical composition stainless steel must be properly heat-treated to secure the best corrosion resistance. With stainless steel of the cutlery type this resistance is lowered considerably by tempering it, after hardening, at temperatures between 500° and 600° C. This has proved a real problem in brazing knife-blades on to metal handles. In service, the blades begin to show signs of corrosion about an inch or so from the bolster where, during the brazing, the temperature is within the above range. Austenitic stainless steel of the " 18–8 ", and similar types, is particularly prone to intercrystalline attack by corrosive liquids after being reheated within the range of temperature, 600° to 900° C. This kind of corrosion is often described as " weld-decay " because it was first observed in welded articles. Actually it has nothing to do with welding and its occurrence in welded structures is merely incidental, due to the fact that the steel, at some distance away from the welded joint, becomes heated within the 600°–900° C range of temperature. Some years ago the position was regarded as so serious that the Air Ministry instituted a standard corrosion test, to which all austenitic stainless steel used in aircraft construction must be subjected. The test is carried out by boiling flat test-pieces of the steel for a period of 72 hours in a solution of copper sulphate and sulphuric acid. At the end of this time the pieces are taken out, washed and cleaned, and then bent through an angle of 180°. If cracks develop during the bending the material is unsatisfactory.

" Weld-decay " is attributed to the deposition of the complex chromium carbide from the austenite, and thus would be expected to be worse the higher the carbon content of the steel. It is for this reason that in the manufacture of the steel every effort is made to keep the carbon content as low as possible. The susceptibility to weld-decay is also lessened considerably by increasing both the chromium and nickel percentages and, nowadays, " 18–8 " is more often " 19–9 ".

To prevent this kind of corrosion entirely, the steel should be heated to about 1100° C., and then cooled in air, preferably by means of an air-blast. With welded structures this is not always possible, and in these cases we have to rely on modified chemical compositions and the addition of other elements. Silicon, tungsten, copper, titanium, or niobium (columbium) are added to austenitic stainless steels solely in order to reduce the susceptibility to weld-decay to an absolute minimum, and therefore obviate the necessity for a final heat-treatment after welding. It will be seen that this problem, which has been a very real one in the past, can be got over quite easily.

Work-hardening.

One other phenomenon must be discussed, i.e. the phenomenon known as " work-hardening ", since it is of particular importance in connexion with the austenitic nickel-chromium corrosion-resisting steels such as " 18–8 ". By work-hardening we mean the increase in hardness which occurs when a piece of the steel in the cold condition is subjected to any degree of deformation such as is involved in the operations of hammering, drilling, pressing, rolling, &c. This increase in hardness by cold-working is responsible for some of the difficulties and troubles encountered in the manipulation of austenitic steels generally and, in particular, of manganese steel.

A simple explanation of the process of work-hardening is difficult, but we hope that the following attempt to explain what happens will be understood by the reader. When a certain stress, less than the so-called " elastic limit ", is applied to steel at normal temperatures it behaves as a perfectly elastic material and returns to its original form and size when the stress is taken away. If stressed beyond the elastic limit, the steel does not return to its original dimensions but acquires a definite amount of permanent deformation or " permanent set ". To illustrate this consider two strips of thin spring steel, one of which has been properly hardened and tempered, the other being in the soft state. Now bend them by means of the hands through the same angle. The treated strip will spring back when the force exerted by the hands is released but the other strip remains bent, i.e. it has been permanently or plastically deformed. During plastic deformation the crystals of which the steel is composed are distorted but it is not a case of mere pulling-out, but rather a process of slipping along the crystal planes of the crystals. Imagine a pile of books sliding over each other when pushed from one side. That is how plastic deformation in steel crystals takes place. But how are we to account for the increased hardness which accompanies this plastic deformation? During the slipping process atoms in the crystal planes are removed from their normal positions. To do this energy must be expended, and in a plastically deformed material this energy is stored up in it. The consequence is that the potential energy of a steel which has been deformed in the cold is greater than that of the soft unworked material. In one way it is like raising a weight to different heights and then allowing it to fall to the ground. When the weight is dislodged from the greatest height, the deeper will be the impression, or dent, formed in the ground. The apparent hardness of the weight thus depends on the height from which it is dropped, i.e. on its potential energy, or energy of position. In a similar way the potential energy and hardness of a cold-deformed steel are proportional to the degree to which it has been deformed or cold-worked: the greater the amount of cold-deformation the harder becomes the steel. Eventually, slip along the crystal planes can no longer occur and if further deformation is attempted the material fractures. To avoid breakage in this way the steel is softened by reheating and cooling from the proper temperature. This gives to the steel its normal structure and further deformation can then be applied without the risk of breakage taking place. With most steels a slow cooling from the softening temperature is essential, but with stainless steel of the " 18–8 " type the best results are obtained by rapid cooling, by means of air-blast or even by quenching in water. Austenitic stainless steel of the " 18–8 " type rapidly work-hardens and frequent softenings are required before it can be reduced by cold-work to the necessary size and shape.

From what has been said it should be evident that a cold-worked metal such as steel is stronger, and more resistant to deformation, than the unworked material. It should be remembered, however, that cold-work may adversely influence the corrosion-resistance of the austenitic nickel-chromium steels.

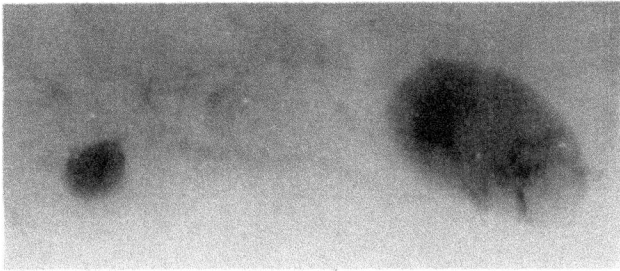

Fig. 39

Fig. 40

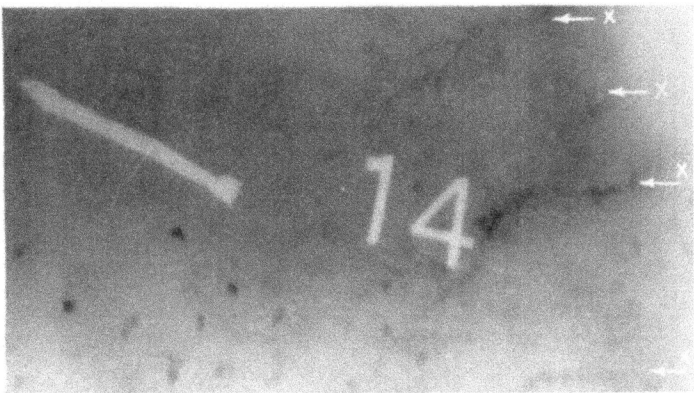

Fig. 41

Facing p. 102

CHAPTER XIII

X-Rays

Electro-magnetic Radiations.

Within recent years X-rays have been employed in metallurgy for two distinct purposes: (*a*) for the investigation of arrangement of the atoms in a piece of metal, i.e. for the determination of its atomic structure, and (*b*) for the detection of internal flaws such as blowholes, segregates, internal cracks, &c. The X-ray has been described as the " all-seeing eye " of science by which the inner structures of substances may be observed.

These remarkable rays were discovered as long ago as 1895 by Professor Röntgen, and appropriately enough, are sometimes known therefore as Röntgen rays. X-rays are now known to be a form of radiation similar in character to radiant heat, light, wireless and other electro-magnetic rays or waves. In these days of wireless transmission and reception it is not too much to expect readers to be familiar with the terms " frequency " and " wave-length ", but their exact meanings may not, perhaps, be fully understood. Let us consider an alternating current such as that usually supplied for lighting and heating. As represented in fig. 37, the voltage of the current undergoes a periodic variation with time, i.e. the voltage passes through zero, rises to a positive maximum, decreases again to zero, then rises to a negative maximum and finally decreases to zero once more, the sequence of changes then being repeated. The distance between A and E is known as a complete cycle or wave-length, and the number of times this cycle is repeated in one second is known as the frequency of the current. For ordinary lighting current this frequency is about 50 cycles per second. Now radiant heat and light consist of electrical waves of a similar kind but their wave-lengths are very much smaller. The essential difference between heat and light rays or waves is, in turn, simply a difference of wave-length: heat waves are longer than light waves and have a much greater penetrating power. Again, the different colours of the rays of the visible spectrum, which together constitute ordinary white light, are due to a difference in wave-length. Thus the wave-length of red light

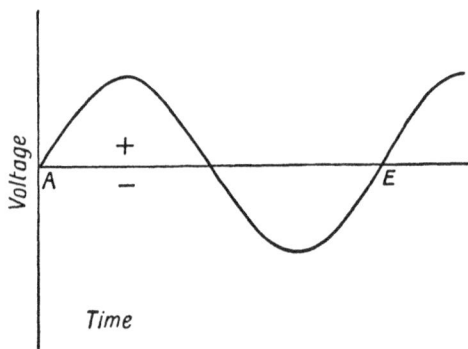

Fig. 37

is longer than that of violet light, and the wave-lengths of the other coloured rays of the visible spectrum, viz. orange, yellow, green, and blue, are intermediate between those for the red and violet.

Infra-red rays, which are invisible to the naked eye, have a longer wave-length than red rays and included among them we have the heat rays. Rays of still longer wave-length are the electro-magnetic rays used in wireless transmission and here, as most readers will be aware, the range of wave-lengths varies from a few hundred feet to several miles. In passing, it is of interest to note that in the modern high-frequency furnaces for the melting of steel and other alloys current

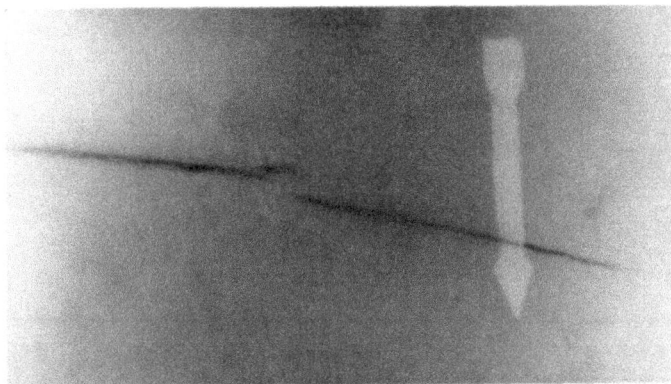

Fig 42

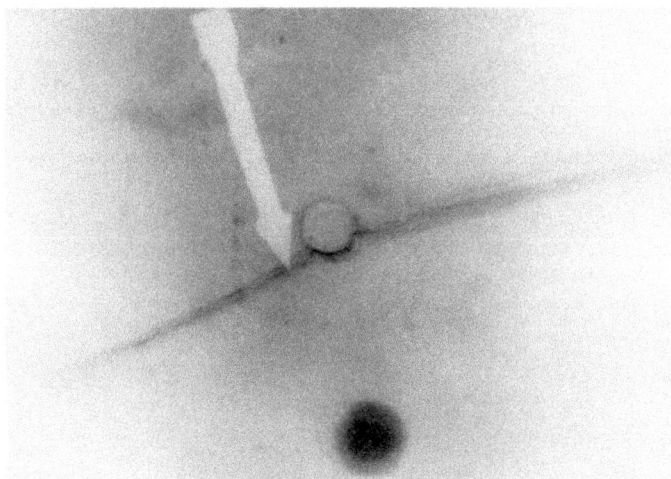

Fig. 43

Facing p. 104

having a wave-length of the same order as that of wireless waves is used.

X-Rays: their Diffraction.

Beyond the violet we have the invisible " ultra-violet " waves or rays of much shorter wave-length. Following upon these we come to the X-ray range consisting of waves of still shorter wave-length. X-rays are therefore electro-magnetic waves of exceedingly short wave-length. Actually, their wave-lengths are of the order of 10^{-8} centimetre, i.e. a tenth of a millionth of a millimetre, and, although they are invisible, they do possess some of the properties of visible light. Thus they can be reflected, refracted or diffracted. In connexion with diffraction the first difficulty which arose was in obtaining a suitable diffraction grating. For ordinary visible light experiments, the diffraction grating consists of a series of fine parallel and equidistant lines ruled on a smooth polished surface of glass or metal so that the distance between each line is of the same order of magnitude as that of the wave-length which is to be measured. Even with the visible light rays this needs several thousands of lines to each centimetre, so that with X-rays of much smaller wave-length the use of a diffraction grating seemed to be entirely out of the question. Von Laue in 1912, however, conceived the happy thought of using crystals, in which the molecules or atoms were regularly spaced, as diffraction gratings, since he had previously come to the conclusion that the distances between their molecules were of about the same order as the wave-lengths of X-rays. Von Laue thus tried the experiment of passing a beam of X-rays through a thin crystal plate and allowing the diffracted rays which passed through to strike a photographic plate, which was then developed and fixed in the usual manner.

Laue Pictures: the Bragg Law.

The Laue pictures thus obtained showed a regular and symmetrical arrangement of dots and from these the exact wave-length of the X-ray could be obtained by mathematical calculation. Once the exact wave-length of the X-ray beam was known, it became a natural consequence to use the X-ray to determine the spacings of the molecules or atoms in other salt crystals and in metals and alloys such as iron and steel. Laue photographs also serve to reveal the effects of cold-work, crystal size, and internal strain in metals. The photographs obtained by the Laue method are not easily interpreted, however, and the greatest advance in our knowledge of crystal structure is due to Sir William Bragg, who realized that X-rays would penetrate the actual surface of the metal, but that, underneath the surface, the surfaces or planes in which the atoms were situated would serve as internal reflecting planes for the X-rays, in much the same way as ordinary light is reflected from a plane polished surface. There is this difference, however; the reflected rays from the different layers interfere with each other and a " fuzzy " photograph is thus obtained unless the relationship

$$2d \sin\theta = n\lambda$$

is fulfilled, where d is the distance between two successive layers of atomic planes, θ is the angle (the glancing angle)

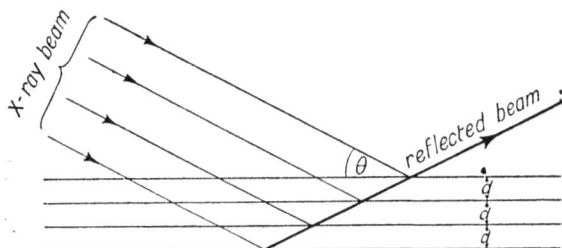

Fig. 38

which the X-ray beam makes with the surface of the alloy under examination, λ is the wave-length of the X-ray, and n is an integer (i.e. a whole number), as represented in fig. 38. This relationship is known as the Bragg law.

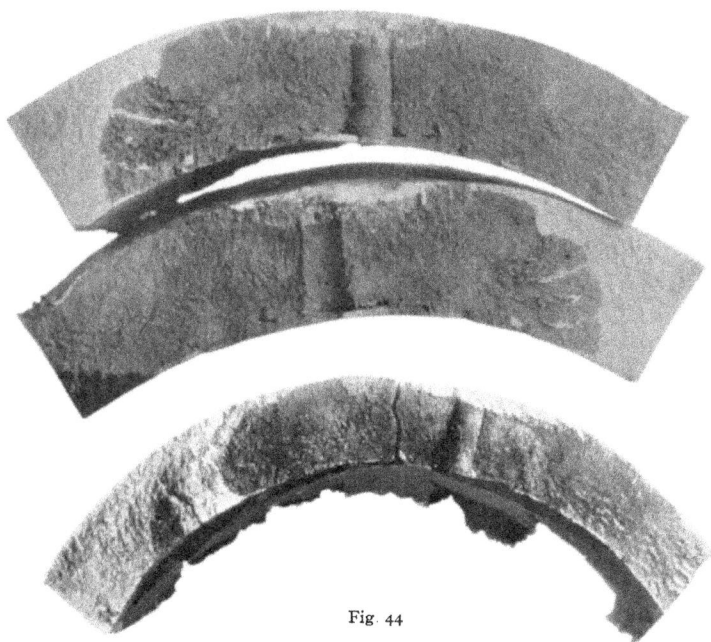

Fig. 44

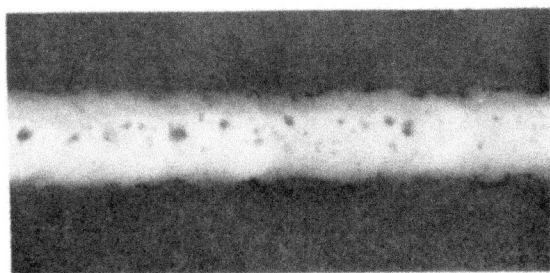

Fig. 45

X-Rays and Crystal Structure.

When these conditions are satisfied each reflected ray reinforces others and a series of sharp bands is then obtained on a photographic film placed in such a position as to receive the reflected rays. The bands or lines on the film are then spaced at definite distances, and from these the distances between the atomic planes may readily be obtained by calculation, i.e. by this means the distances between the atoms may be ascertained. It was in this way that the fundamental structures of alpha and gamma iron, referred in to Chapter IV, were determined, where, it will be remembered, alpha iron was considered to possess a body-centred space-lattice and gamma iron a face-centred cubic lattice. Other metals and their alloys have been examined by this method, which is truly ultra-microscopic, and much fundamental information regarding their structures has been obtained thereby.

X-Rays and Interior Pictures of Metals.

There is, however, another and perhaps still more important practical application of X-rays to industrial metallurgical problems. By using X-rays of greater penetrating power it is possible to obtain interior pictures of metals and alloys such as iron and steel. These are obtained in much the same way as X-ray pictures of the human body are obtained by the medical and surgical radiologists. In such cases, the resulting pictures are really shadowgraphs produced as a consequence of the fact that bones, foreign bodies and certain organs of the human body are much more opaque to X-rays than is the flesh. In a similar fashion voids, in metals and alloys, such as blowholes or gas cavities, internal cracks, included non-metallic matter, &c., allow the X-rays to pass through them more easily than through the metal itself. These defects are then revealed on the X-ray plate or film as dark patches as shown in fig. 39. In many cases the outside surfaces of such articles as castings may be perfect and it is only when they are (i) sectioned or (ii) subjected to X-ray examination that internal fissures, &c., are revealed (see figs. 39 and 40). In the first method the article must be destroyed, so that there is at once an obvious advantage associated with the X-ray test, which is non-destructive. The X-ray has proved particularly useful in the examination of welds.

Fig. 39 is a radiograph of a section of a steel casting for high-pressure steam line, showing gas cavities and an extended area of shrinkage.

Fig. 40 is a photograph showing section taken through both gas cavities and one of the shrinkage areas.

Fig. 41 is a radiograph of another section of the same fitting showing shrinkages near the flange; apparently square nails were used for chilling, as indicated radiographically at X.

Figs. 42 and 43 are reproductions from radiographs of two castings for a high-pressure steam line in a power plant. These cracks appear in the same place, at the chaplet, indicating a faulty casting technique.

Fig. 44 shows photographs of the two castings shown in figs. 42 and 43, after sectioning through the cracks (dark areas).

Fig. 45 is a radiograph showing cavities in a weld metal. This weld seemed an excellent one, and would have passed ordinary inspection.

From the above diagrams it will be seen that the X-ray constitutes a most valuable means for the revelation of internal defects in such materials as iron and steel and, as already pointed out, the test is non-destructive in character, which is perhaps its greatest advantage. At the moment, unfortunately, the main difficulty lies in generating X-rays of sufficient penetrating power, and articles of more than a few inches in thickness cannot be satisfactorily examined by this method.

Electron-Diffraction.

Within recent years the *electron-diffraction* method for the examination of metals has been developed. In this case a beam of electrons is allowed to impinge at an angle on the surface of the material under test and the deflected beams can be registered on a special photographic plate or film as in the case of X-rays. There is this important difference, however. The electron beam does not penetrate the surface layers like X-rays, and the electron-diffraction pattern thus indicates the structure of the actual surface. This method of examination is still in its infancy but future results may play a most important part in connexion with our knowledge of the properties of metals, since it is becoming increasingly recognized that the properties of materials such as stainless steel are largely the properties of their surface layers. All the other methods indicate the structure of the metal beneath the surface. Here, then, is another useful and important metallurgical tool for the examination of metal structures.

TABLE I

BRINELL'S HARDNESS NUMBERS

Diameter of Steel Ball .. 10 mm.
Pressure .. .. .. 3000 kg.

Diameter of Ball Impression	Hardness Number	Calculated Tonnage	Diameter of Ball Impression	Hardness Number	Calculated Tonnage	Diameter of Ball Impression	Hardness Number	Calculated Tonnage
mm.			mm.			mm.		
2	946	206	3·70	269	59	5·35	124	28·5
2·05	898	196	3·75	262	57	5·40	121	28
2·10	857	187	3·80	255	55	5·45	118	27
2·15	817	178	3·85	248	54	5·50	116	26·5
2·20	782	171	3·90	241	52	5·55	114	26
2·25	744	162	3·95	235	51	5·60	112	25·5
2·30	713	155				5·65	109	25
2·35	683	149	4	228	50	5·70	107	24·5
2·40	652	142	4·05	223	49	5·75	105	24
2·45	627	136	4·10	217	47	5·80	103	23·5
2·50	600	131	4·15	212	46	5·85	101	23
2·55	578	126	4·20	207	45	5·90	99	22·75
2·60	555	121	4·25	202	44	5·95	97	22·5
2·65	532	116	4·30	196	43			
2·70	512	112	4·35	192	42	6	95	22
2·75	495	108	4·40	187	41	6·05	94	21·5
2·80	477	104	4·45	183	40	6·10	92	21
2·85	460	100	4·50	179	39·5	6·15	90	20·75
2·90	444	97	4·55	174	39	6·20	89	20·5
2·95	430	94	4·60	170	38·5	6·25	87	20
			4·65	166	38	6·30	86	19·75
3	418	91	4·70	163	37·5	6·35	84	19·25
3·05	402	88	4·75	159	36·5	6·40	82	19
3·10	387	84	4·80	156	36	6·45	81	18·75
3·15	375	82	4·85	153	35	6·50	80	18·5
3·20	364	79	4·90	149	34	6·55	79	18·25
3·25	351	76	4·95	146	33·5	6·60	77	17·75
3·30	340	74				6·65	76	17·5
3·35	332	72	5	143	33	6·70	74	17
3·40	321	70	5·05	140	32	6·75	73	16·75
3·45	311	68	5·10	137	31·5	6·80	71·5	16·5
3·50	302	66	5·15	134	31	6·85	70	16·25
3·55	293	64	5·20	131	30	6·90	69	16
3·60	286	62	5·25	128	29·5	6·95	68	15·75
3·65	277	60	5·30	126	29			

TABLE II

CONVERSION TABLE OF VARIOUS HARDNESS NUMBERS

This table applies only to chemically and mechanically uniform steels containing carbon, chromium, nickel, vanadium, molybdenum, silicon and manganese, none of the alloys exceeding 4 per cent. No cold-worked steels are included. All testing methods used were those recommended by the manufacturers of the respective instruments.

Vickers or Firth	Brinell Diameter mm.	Brinell Standard Ball	Brinell Tungsten Carbide Ball	Rockwell C 150 kg.	Rockwell D 100 kg.	Rockwell A 60 kg.	Rockwell B $\frac{1}{16}$-in. Ball	Rockwell E $\frac{1}{8}$-in. Ball	30-N	30-T	Durometer	Monotron Cons. Dia.	Monotron Diam. Brin.	Drop Hardness mm. kg	Herbert Pendulum Time S	Herbert Pendulum Time D	Scleroscope	Duroskope	Moh
1224	2·20	780	872	72	78	84			87			28	130	1200		90	64	106		
1116	2·25	745	840	70	77	83			86			31	122	1130		85	59	102		8·5
1022	2·30	712	812	68	75	82			84			32	111	1030		80	56	98		
941	2·35	682	794	66	74	81			83			34	103	956		76	52	94		
868	2·40	653	760	64	72	80			81			37	96	894		72	49	91	54	
804	2·45	627	724	62	71	79			79			40	91	850		67	47	87	53	8·0
746	2·50	601	682	60	70	78			78			43	86	804		63	45	84	52	
694	2·55	578	646	58	68	78			76			45	82	767	1400	60	42	81	51	
650	2·60	555	614	56	67	77			74			46	78	727	1300	56	40	78	50	7·5
606	2·65	534	578	54	65	76			72			48	74	690	1225	53	38	76	49	
587	2·70	514	555	52	64	75			70			49	71	660	1160	51	37	73	49	
551	2·75	495	525	50	63	74			69			50	68	630	1095	48	36	71	48	
534	2·80	477	514	49	62	74			68			51	66	610	1050	47	35	68	48	
502	2·85	461	477	48	61	73			67			52	63	586	1005	44	34	66	47	
474	2·90	444	460	46	60	73			66			53	61	566	950	41	33	64	46	7·0
460	2·95	429	432	45	59	72			65			55	59	548	910	30	32	62	46	
435	3·00	415	418	43	58	72			64			56	57	530	880	37	30	61	45	
423	3·05	401	401	42	57	71			63			58	55	510	840	35	29	59	45	
401	3·10	388	388	41	56	71			62			59	53	490	810	34	28	57	44	
390	3·15	375	375	40	56	70			61			60	51	471	780	33	27	56	44	6·5
380	3·20	363	364	39	55	70			60			62	50	462	750	32	26	54	43	
361	3·25	352	352	38	54	69			59			63	48	453	725	30	26	53	43	
344	3·30	341	341	36	53	68			58			64	47	433	700	29	25	51	42	
334	3·35	331	330	35	52	67			57			65	45	414	675	28	24	50	42	
320	3·40	321	321	33	50	67			56			66	44	408	650	27	24	49	41	
311	3·45	311	311	32	50	66			55			68	42	390	630	26	23	47	41	
303	3·50	302	302	31	49	66			54			69	41	380	610	24	23	46	41	6·0
292	3·55	293		30	48	65			53			70	40	370	590	24	22	45	40	
285	3·60	285		29	47	65			52			71	39	360	570	23	22	44	40	
278	3·65	277		28	47	64			51			72	38	350	550	23	21	43	39	
270	3·70	269		27	46	64			50			73	37	340	535	22	21	42	39	
261	3·75	262		26	45	63			49			75	36	331	520	22	20	41	38	
255	3·80	255		25	45	63			48			76	35	322	505	21	20	40	38	
249	3·85	248		24	44	62			47			77	34	313	490	21	19	39	37	5·5
240	3·90	241		23	43	62	102		46	85	78	33	304	475	20	19	38	37	
235	3·95	235		21	42	61	101		45	84	79	32	295	460	20	19	37	37	
228	4·00	229		20	41	61	100		44	83	81	31	286	450	19	18	36	86	
222	4·05	223		19	40	60	99		43	82	82	30	277	440	19	18	35	36	

[Continued]

Vickers or Firth	Brinell Diameter mm.	Brinell Standard Ball	Brinell Tungsten Carbide Ball	Rockwell C 150 kg.	Rockwell D 100 kg.	Rockwell A 60 kg.	Rockwell B 1/16-in. Ball	Rockwell E 1/8-in. Ball	30-N	30-T	Durometer	Monotron Cons. Dia.	Monotron Diam. Brin.	Drop Hardness mm. kg.	Herbert Pendulum Time S	Herbert Pendulum Time D	Scleroscope	Duroskope	Moh
217	4·10	217		17	39	60	98	110	42	82	83	29	268	430	19	18	34	36	
213	4·15	212		15	38	59	97	110	40	81	84	29	268	415	18	17	34	35	
208	4·20	207		14	37	59	95	110	39	81	86	28	259	405	18	17	33	35	
201	4·25	201		13	37	58	94	109	38	80	87	28	259	395	18	17	32	35	
197	4·30	197		12	36	58	93	109	37	79	88	27	250	385	17	16	31	34	
192	4·35	192		11	35	57	92	108	36	78	89	26		375	17	16	30	34	5·0
186	4·40	187		9	34	57	91	108	35	78	90	26		365	17	16	30	34	
183	4·45	183		8	34	56	90	108	34	77	92	25		360	17	16	29	33	
178	4·50	179		7	33	56	90	107	33	77	93	25		350	17	16	29	33	
174	4·55	174		6	33	55	89	107	32	76	94	24		340	16	15	28	33	
171	4·60	170		4	32	55	88	106	31	76	95	23		335	16	15	28	32	
166	4·65	167		3	32	54	87	106	30	75	96	23		330	16	15	27	32	
162	4·70	163		2	31	53	86	105	29	74	97	22		320	16	15	27	32	
159	4·75	159		1	31	53	85	105	28	73	99	22		315	16	15	26	31	
155	4·80	156		0	30	52	84	104	27	73	100	21		310	16	15	26	31	
152	4·85	152					83	104		72	102	21		300	16	15	25	31	
149	4·90	149					82	103		71	104	20		295	15	14	24	30	4·5
146	4·95	146					81	103		71	105	20		290	15	14	24	30	
143	5·00	143					80	102		70	106	19		285	15	14	24	30	
140	5·05	140					79	102		69	108	19		280	15	14	23	29	
138	5·10	137					78	101		68	109	19		275	15	14	23		
134	5·15	134					77	101		68	111	18		270	15	14	23		
131	5·20	131					76	100		67	113	18		265	14	13	22		
129	5·25	128					75	100		67	115	18		260	14	13	22		
127	5·30	126					74	99		66	116	17		255	14	13	22		
123	5·35	123					73	99		66	117	17		250	14	13	21		
121	5·40	121					72	98		65	119	17		245	14	13	21		
118	5·45	118					71	98		64	121	16		240	14	13	21		
116	5·50	116					70	97		63	122	16		235			20		
115	5·55	114					68	97		63	123	16		233			20		
113	5·60	111					67	96		62	124	15		230			20		
110	5·65	110					66	95		61	126	15		228			20		
109	5·70	109					65	95		60	127	15		225			19		
108	5·75	107					64	94		59	128	14		223			19		

INDEX

*9 7 8 1 4 7 9 4 4 5 6 3 9 *